REBUILDING THE GE HOUSE JACK BLEW DOWN

40 Years of chutzpah and sick humor at GE

Mary Kuykendall

Copyright © 2015 Mary Kuykendall
All rights reserved.

ISBN: 1514304740
ISBN 13: 9781514304747

SYNOPSIS

A non-fiction work, it is part memoir and part business to show what it is like to have lived and worked through the eras of paternalism and today's materialism or as some business writers producing books about big business today have called it, the era of CEO greed. Most books about business, even the ones exposing greed and corruption, are written from an academic or business reporting viewpoint. This personal story takes one through the years of union power in the 60s and 70s and how big business got the upper hand in the 80s and 90s starting in the Reagan era. It also shows the impact of the 1964 Civil Rights movement and the 1972 Equal Employment Opportunity Class Action Suit rulings on GE culture.

Most of all, it documents the power of how one CEO can harvest the company for the benefit of himself and his Gold Collars, a new management class he introduced in the company which sits atop the traditional blue, pink (women) and white collar system. For nearly 30 years now, these groups have been asked to put in a "fair day's work for a fair day's pay," an era which has led to stagnant pay for them. In the paternalistic days, top management would receive 15 times the pay of the white collar professional, today it is more than 500 times. Now those tens of thousands of blue and white collar employees who were unnecessarily laid off are asking the Gold Collars to put in a fair day's work for a fair day's pay. Harvesting of GE included selling or moving 98 U.S. plants, most of them offshore and getting rid of over 180,000 U.S. middle class jobs. Harvesting also included selling advanced technology, not investing in research and development, limiting plant and equipment expenditures, closing product develop labs and forcing local and state governments to reduce tax rate assessments as well as contribute to plant and equipment investments by threatening to move. How could one CEO accomplish all of this in the world's most diversified company?

This account tells you how Jack Welch did it. His successor is now finding it is a lot harder to grow a company then harvest it.

CHAPTERS

A Rocky Start	1
Those Early GE News Years	15
Making Schenectady Competitive	30
Motor Years	51
No Thanks	68
Fitting In at Headquarters	80
My Dream Come True Job	101
The Houses Jack Harvested and Blew Down	123
How the Gold Collars Sold Us Out	148
Sick Humor- Dealing with the Insanity	173
Whatever Happened to a Fair Day's Work for a Fair Day's Pay?	197
Being My Own Boss, What Fun	207
GE Today, Good Luck Immelt	220

DEDICATION

To the tens of thousands of GE blue and white collar employees who unnecessarily lost their jobs.

A ROCKY START

Looking back on a chaotic writing career at GE, I realize I could be my own worst enemy because of who I was. At West Virginia University, between my junior and senior year, I was an intern at the Wheeling News-Register where I was steeped in Fourth Estate rules of who, what, where, when, how and why. The summer before I had been a carhop on roller skates at a huge drive-in near the state capitol in Charleston where I was fired because I dumped milkshakes on two sons of a state senator who had grabbed my money belt. From the age of 10, I had worked sugar corn fields and fruit orchards from sunup to sundown, most without rest except at the peach orchard which was owned by an evangelist who sent you off to his orchard chapel if you cursed. In fact, I was about to run off with a cursing stock car driver I met in that chapel when my older sister told me she had enrolled me in WVU. She graduated four years ahead of me. She got a job at Union Carbide and was paying off the mortgage on the farm which had gotten her through college. Now she wanted to help me.

She was the brain in the family, having graduated cum laude in chemistry and math while working in freshman chemistry labs as an assistant. My interests lay in Journalism which was particularly attractive to me when I met what we called ourselves then -- beatniks, the precursors to hippies. I even met Ferlenghetti when he visited the campus. But it was a Journalism professor by the name of Paul Atkins, who oversaw the daily Athenaeum student newspaper, who inspired us to resist unfairness. He resented the fact the fraternities and sororities were so powerful they even had control of the University yearbook. In just two years, we independents could run for student office, sit on the 50-yard-line at football games and even have our own float in the homecoming parade. One of these independents was Basketballer Jerry West who made all-American his sophomore year. When he became well-known, the top fraternity sought his membership but he turned them down. He was as smart as he was talented. No one got paid $50 for taking a test for him.

While I would like to say this was one way I was able to supplement my sister's handouts for room and board and tuition, only those with 3.5 averages were tapped for athletic help. Not only was I in a party school but part of the new Journalism curriculum was to take courses in various colleges making up the university. The aim was to give us a working knowledge of all professions. I quickly chose astronomy being an admirer of the night sky. Soon I found myself on top of the Physics building with a spectrometer trying to measure the distance to stars. I realized I was about to become the J-School's failure in that experiment when the professor mentioned his desire to have a portable planetarium to help him recruit high school students in the summer. I not only wrote articles for our student newspaper but sent press releases to media throughout the state seeking financial support. Fortunately, he also assigned essays in his class so my paper on constellations not being real so he would not think I was into astrology resulted in my getting a C in the course. It also resulted in my getting a key to the university planetarium where our Athenaeum gang would gather for a party under the revolving stars when classes were not being held there. As the host, it was not a BYO (*Bring Your Own*) for me.

But unlike my sister who worked as an intern in chemistry labs, I did find other less-skilled ways to help with the tuition, room and board. My first job was stacking (*and reading*) newspapers that came in from all over the country to the Journalism School. I also worked in the kitchen serving 300 women, mostly using a paddle to stir the lima beans or mashed potatoes. The waiters were uniform-clad fraternity men who soon learned how to properly set a table as instructed by the Dean of women who sat at the head table. No one left until she did so I was happy not to have to formally sit there for hours until she got up.

I did lose this kitchen job when she discovered it was me that had been making snowballs out of leftover mashed potatoes with one careening out the door when a waiter came in with a tray. Other money came from passing out those little packs of four cigarettes for Viceroy and Marlborough. I lost this job when they found out I had started smoking so all the inventory was not getting passed out as planned.

But the best money came from a job I had putting up the mail in the women's dormitory. Rules were strict then with a lockdown to your room at 7 p.m. or you could stay out until 9 p.m. if you made above a 3.0 average. Realizing there was a lot of rich Chestnut Hill girls from Philadelphia that were more liberated, I had copies made of the postmaster's key to the only door without a buzzer and sold them.

So unlike my sister, I did not graduate cum laude. But the Dean of Women did note that I had graduated with more restrictions for breaking house rules than any other woman during her reign. I think I made her the maddest when I refused to get up when she came into my room. I quickly sat on two coke bottles on the bed which would have meant two nights more of lockup for an untidy room versus one night for not properly addressing her by getting up.

But she did do me a favor when I inadvertently helped her get police action against the panty raids the fraternities were famous for in those days. I had a boyfriend in the ROTC who gave me some M-80s. He, too, chafed at how independents were treated as second-class citizens because they could not afford affiliation. So when the frat rats were charging at the front double door framed by four massive pillars, I was ready for them. I was in the balcony above the entrance armed with what I thought were just firecrackers. They turned out to be powerful enough to draw the attention of the Morgantown police. Prior to that they just ignored calls from the Dean claiming "boys were just being boys"….that no one was harmed…that they only raided the panty drawers. But when the call came in about explosions leaving a few holes in the ground, lots of yelling and reports of a busted camera carried by one of the raiders, they were on their way. The dean called all 300 women to the dining room where she announced (*looking directly at me*) the police would be looking for what she thought was dynamite. It didn't take me long to scurry back to my room to break up the rest of the M-80s I had and flush them down the toilet.

It was the boyfriend in the ROTC who suggested I join the marines as a way to finance my way through college. They not only accepted me but promised I would be a public information officer when I graduated. I had a friend from the dorm who used to help me raid the soda machines in the dormitory section housing girls with money before they left for sorority houses their sophomore year. She flunked chemistry but we used the laboratory tongs and tubes she borrowed from the lab to suck soda out of the bottles which were strung along a metal rail making the caps available for easy removal. The Marine recruiter said she could be my jeep driver so we both enlisted together. When I told my sister about it, she just hung up and called the recruiter. So my friend and I were only Marines for two weeks after my sister submitted our birth certificates *(my friend ended up as head of the guards at a West Virginia prison for women)*.

Oddly enough, my best memories of college are the jobs I took to help pay for it. The best summer job I had was one arranged by the School of Journalism as an intern between my junior and senior years. In fact, I was due to go back to the Wheeling News Register when I graduated to become a staff reporter at $70 a week. But this changed during several School of Journalism graduation parties *(we hated leaving each other)* when employment ads were being read aloud. The one from GE was particularly intriguing because it promised opportunities unlimited in Schenectady New York, the mother ship of GE with 28,000 employees in the city that lights and hauls the world.

It was the bit about being the birthplace of public radio and TV that really got us going with my fellow students voting that I could show them a photo of a football game I took that made the AP wire. Our sports writer for the Athenaeum called in sick right before kick-off time and since I was there stacking newspapers, the Professor sent me to the game with the Speed Graphic camera. After nearly being crushed by a sideline pass, I decided to go to the other end of the field where I found the university mascot, a Mountaineer with a muzzle loader. I persuaded him to climb up the goalpost so I could get a shot of him aiming his shotgun at the scoreboard showing us leading. As chance would have it, an interception occurred at the other end and soon we had the whole field rushing toward us. It was then the wooden goalpost gave way with all of us in a pile in the end field. But I did get the photo. And AP loved it since it fit in with the stereotype public image of West Virginia.

Even Norman Rockwell's Saturday Evening Post got into making fun of West Virginia when they offered to cover the homecoming game if the queen would be barefoot when crowned. They had already featured our basketball star, Jerry West, as the Hick from Cabin Creek, West Virginia. Despite our best efforts, we seem to lose even when winning games. When it was announced the games would be filmed in color, Coach Art Lewis had the West Virginia football field dyed green to show off the field. The dye proved to be color fast on the uniforms with both sides turning green by halftime.

After much teasing about my sports photography, our Athenaeum gang decided they should help me fill out an application for "unlimited opportunities" in Schenectady. We accomplished this at a party in the planetarium. To this day, I am amazed that I got a call from GE asking me to meet the editor of the GE News in New York City. The letter went on about my unlimited capacity for adult refreshments *(I won the coveted School of J Bladder Award in 1959)* as well as my health, having been kicked off the soccer team because I sent both the sorority girl and the ball into the cage. But it was the health section that may have been the key to an interview. The professor, upset that I was not continuing his passion for the Fourth Estate at the Wheeling News Register, noted that I was fairly good looking and companies did not want to hire somebody who looked marriageable. He suggested I add that I had a hysterectomy.

When GE said it would pay my way to the big city, I set up another interview with a brokerage house looking for a communicator. But I never got to that interview scheduled three hours later. When I got off the Eastern airlines plane at what was then Idlewild airport (*JFK was still alive*) I met the editor, Al Denniston, who looked amazed when he saw me wearing a picture hat. My mother, who had always tried to make southern ladies out of her daughters, insisted I wear it. Considering the letter I had written about my "athletic" abilities, I decided it might be a good idea.

What I really had going for me, however, was a huge book of clippings from articles I had written for the Wheeling News-Register and those in our daily newspaper at the University, including being the editorial writer. He had been impressed -- not so much by my interviews with the nation's leading faith healer at the time, Kathryn Kuhlman -- but the fact that the subscriptions had jumped nearly 10,000 as a result. While I was reporting on every word the faith healer said, a co-reporter was interviewing church leaders and other doubters of her ability. Perhaps Denniston realized I would be good for the world of propaganda even though I expressed my view that in the end it was found Kathryn was a fraud because her simple home was filled with old master paintings. He was less enthralled as to how I got this interview when other reporters had failed. I noticed that all you had to do to get on stage with her at the old WWVA Virginia Theater in Wheeling was to cough or come in with a cane or brace. I coughed and soon found myself on stage. I was soon trembling but not just because of her towering figure topped with penetrating green eyes and red hair. There were also hundreds of people looking at me. She soon figured me out and announced I was not a believer. Kathryn had one of her twelve disciples escort me off the stage as she called on other sickly looking or sounding people. They let me wait for her in her dressing room at which time I confessed she was my first big interview assignment from the editor. Maybe she felt sorry for me, but in any case she agreed to be interviewed if she saw and approved of what I wrote before it was printed. The editor, Harry Hamm, was happy to oblige realizing all he had to do was have other reporters interview medical and religious leaders to have a great "for" and "against" controversy to boost readership. Hamm even met Kathryn when she came in to thank him for the coverage. I still laugh about his comment after she left noting she was quite a step up from the last religious nut we had in Wheeling who was selling autographed pictures of Jesus Christ.

But the interview that really impressed GE's Denniston was the one with Jackie Kennedy. It was the summer of 1959 and Senator Jack Kennedy was doing battle with Hubert Humphrey in the Democratic primary race. Jack's choosing of West Virginia as his kickoff primary had national attention because it would test as to whether a Catholic could win in a state with less than 2% Catholics. I was assigned to Al Molnar, a veteran reporter and political observer, who noted he thought the Kennedy's had made a brilliant choice in choosing West Virginia to take on the Catholic issue. In stating the nation not only branded W.Va. as a bunch of hillbillies but also prejudiced against Catholics, Molnar pointed out the other 98% of West Virginians were used to hundreds of wild religious nuts roaming around the state -- and that the only difference to them was the Catholics had better tents. Furthermore, he declared Joe Kennedy had lots of money and having him campaign as Eleanor Roosevelt's friend made him a shoo-in because she had done so much for West Virginia coal miners and schools.

As Molnar was getting ready to leave to cover a talk Jack was making at the Eagles Club across the river in Bellaire, Ohio, the Mayor's office in Bellaire called to say that Jackie Kennedy was now coming in at the nearby Wheeling, WV airport to meet Jack.

Molnar briefed me as best as he could as he dropped me off at the Wheeling airport, telling me about the Kennedy clan including personal information such as the number of children Ethel had and that Jackie had been named one of the ten best dressed women in the world. I immediately felt I wasn't up to the job, certainly not in an outfit I had since high school. I was conscious of the large safety pin I had used to replace the button which held up the skirt. I had not taken the time to sew it back and told myself that no one would see it under my long pullover blouse. Molnar just laughed and assured me that Jackie would not be paying any attention to me and if she did, it would be with compassion.

As he dropped me off, it hit me that I did not have a way to follow Jackie. We only had one press car. But Molnar was already giving me a $5 bill to buy the Mayor's wife a beer at the airport bar as they waited for Jackie – "that will get you in the front car with her," he said. Then he admonished me not to forget my assignment. I was to try to get Jackie to announce that they were going to accept the presidential nomination. This would be a scoop for the Wheeling News-Register.

When I met the Mayor's wife and her entourage, I felt better about my appearance. They had obviously just left the bowling alley where they had been in the midst of a tournament. They were getting rid of their socks and bowling shoes and putting on heels and hats. As it turns out, all our hurrying had been for naught because Jackie's plane was late. I had about 50 cents left of the $5 bill when it came in and we all rushed out to greet her.

When Jackie stepped off the plane, she did indeed look like the best dressed woman in the world. We had five cars lined up for what the Mayor's wife called her cavalcade. It was a sweltering hot day. It looked like everyone had been promised a ride in the lead convertible as they tried to get in. It was then I spotted an opportunity. I noticed her precarious pillbox hat and asked Jackie if she would be more comfortable in the hardtop car behind the Mayor's car. She immediately followed me out of the backseat and into the car behind me.

She did express surprise that I was a reporter at the age of 17 but she was courteous when she found out I was just a summer intern. I began the interview by asking her how she liked Washington, DC and how exciting it must be there with all the social life, not realizing until years later her social life was not all that good with the wandering Jack. She answered everything with a smile and breathy voice. I became completely enchanted with her as she made me feel comfortable by thanking me for getting her into the hardtop car and encouraging me in my journalism career.

As we were heading down Rt. 40, a double lane highway taking you into Wheeling, a car pulled up in the fast lane beside us with several women dressed in their bowling clothes with their team's name on their T-shirts. The one in front leaned out of her window to ask our driver if we had gotten rid of Jackie yet... that the tournament was being held up. She tried to shush her by pointing to Jackie in the back beside me. But Jackie heard it, and just as quietly, asked me about the bowling game after the car drove on ahead. I gave her the details and even added an apology, explaining that no one expected her to be in this old hardtop car.

Jackie graciously expressed her thanks to everyone for coming to get her and announced that she would do everything she could to speed up their trip so the tournament could go on. When we got to the gift store of the Imperial Glass Company, Jackie asked not to take the tour because she was so late coming in to the airport and that Jack was waiting for her. She also added how nice the welcoming committee had been to wait for her. She then impressed the plant managers by asking them several technical questions about how the glass was made. She revealed a genuine knowledge of glass commenting on several pieces that were not only equal to but better than some she had seen in Venice.

When we got to the Eagles Club, Jackie again thanked the Mayor's wife and her committee. She told them she was sure I could show her inside the club without any trouble. She then wished them a winning tournament and off we went into the club. Jack was in the middle of his talk so Jackie asked that we sit in the back rather than disrupt things by going to the front where she was supposed to be seated. The room temperature was nearly as sweltering as it was outside. I asked Jackie if she would care for a drink. About half way up the side of the room was a barrel containing Iron City beer from Pittsburgh on ice plus a table with iced tea. Those around us were having the beer which she requested and I promptly returned one for each of us. I proudly opened both cans by producing from my purse my beer opener which we called a church key in those days.

During Jack Kennedy's talk I kept reminding myself of my mission from Molnar. When Jack discussed programs he would sponsor to help West Virginia miners, the applause was heavy. It was then I commented on how much everyone seemed to like him and asked if she going to go with him on the presidential campaign trail and she said "yes." That was our headline in the newspaper the next day proclaiming that Jack was running for president and we were the first ones to announce it. *(Years later I would see myself in Photographer <u>Mark Shaw's book, "The John F. Kennedys</u>" with the mayor's wife and her entourage at the Glass Company.)*

But back to the job interview with GE News Editor Al Denniston. When Denniston finished reading some of the articles in my huge book of Athenaeum and Wheeling News Register clippings at the airport, he suggested that he show me New York City because I had never been there. I quickly agreed and put the book in a nearby locker.

After a visit to the Empire State Building, he suggested we have lunch at the Rainbow Room for another view of the city. It was here I began ordering drinks I had heard about but never had in a dry state where the only drinks were served in veteran's clubs. Even there you got your drink by ordering a 7up and lifting the candle off the bottle top which contained Seagram's Seven and not shine if you were lucky. Perhaps it was this hard-core experience that enabled me to consume a Manhattan, then a Martini, a Singapore Sling, a Black Russian, a Stinger and so on. I continued going down the list of drinks to Al's amazement. Fortunately, he had a train to meet to get back to Schenectady. He paid the cab driver to take me to the airport and rushed off. My problem began when the driver asked me which airport. I decided I couldn't have gotten that far away and told him the nearest one. When we got to LaGuardia, I found Eastern Airlines but my locker key to get my book of clippings would not work. The agents couldn't figure out why it wouldn't open either. It had the same number. So a maintenance person was called to blast it open at which time we found nothing.

It was then the agent asked to see my ticket with the discovery that I was at the wrong airport. Needless to say I missed the flight. When I got to Washington DC I missed the train from there to West Virginia. When I walked into our farmhouse a day later, my mother told me that GE had called and I had the job. They would pay me $20 more a week than what I had been offered by the Wheeling News-Register. Then she asked me where my hat was. I did not tell her how it had sailed off the top of the NBC building from the balcony of the Rainbow Room. Nor did I tell her I missed the other interview.

When I arrived by train in Schenectady *(as in most eastern industrialized U.S. cities, your first view is a junkyard)* I wondered what I had gotten myself into. But soon I was making my way to the local YWCA where I knew women could get a room. My experience in Wheeling had shown me that a single woman looking for a room was highly suspect of turning it into a red light locale.

When I began the job employment at the plant was at 28,000. It was a city within a city. The newest building, the world's largest manufacturing building covered over 10 acres. It was number 273 because that was how many buildings had been built since Edison moved his dynamo plant out of New York City in 1891. The plant was vertically integrated. Everything needed to build a turbine, motor or generator was there. I will never forget the smell of cherry, walnut and mahogany as patternmakers carved the shape of turbine parts into the wooden molds. I felt like I was in a Dante's inferno scene in the iron and steel foundries where hot molten metal pours were made to make different castings of all the parts. Production and assembly was also supported by the wire mill, tool and die shops, insulating materials and blueprint operations.

Each product had its testing facility to assure it operated with the precision of a Swiss watch as advertised. In the product development and testing labs, the working environment of vulnerable turning parts such as turbine blades would be simulated to show the effect of being blasted by years of hot steam traveling thousands of miles an hour.

The AC and DC motor and generator business was just as impressive. In Bldg. 60, huge hydro-generators the size of circus tents were being built to generate electricity from water dammed up in the Midwest.

Not everything was huge. The intricate all-white clean room in Bldg. 169 where vacuum tubes were built was just as interesting. The plant had its own power plant to provide electricity, its own railroad and plant bus system, a fire station, a couple of cafeterias, a medical clinic, an employee store with appliances and a huge athletic facility with indoor and outdoor sports.

I was mesmerized by the huge plant and its history. Thomas Alva Edison and electrical pioneers like the brilliant mathematician Charles Proteus Steinmetz gave birth to the Electrical Age. They gave Schenectady the title of "the city that lights and hauls the world." Edison's two original buildings were still there. The world's first steam turbine – a 5000 kilowatt vertical unit installed in 1903 at the Edison Fisk Street Station in Chicago was now a monument in the plant – a salute to units that were now being built to produce over 1,000,000 kilowatts. Not far from it stood GE's first gas turbine which was built in 1949 for Oklahoma Gas and Electric.

Plaques around the plant verified Schenectady as the birthplace of many consumer products such as the first practical hermetically-sealed home refrigerator developed by Christen Steenstrup. Many of those refrigerators, called the Monitor Top because the compressor looked like the gun turret of the Civil War Monitor, are still in operation as are many of the early steam and gas turbines. I thought it was fitting that one of Edison's original buildings was being used now to develop robotic equipment.

I loved exploring every part of the plant with the GE News photographers. In addition to Denniston, there was an assistant editor. My job began as communication specialist. One of my first assignments was to solicit answers from employees on suggested questions which would make the company and employees more competitive. I immediately jumped into doing that feature since it got me into all the buildings.

My enthusiasm did take a dive, however, when I met Denniston's boss, Hal Reed, who obviously had acquired the stereotype view of West Virginia. He lowered my salary by $10 a week noting it was $10 more a week than I had been offered at the Wheeling News Register. He told me I should be grateful to have a man's job in big business. Naturally I was upset but at the same time I was determined to show him I was certainly capable of doing a man's job. Besides, it was too late to go back to Wheeling. That job was gone. I was down to my last dollar. I knew that my father was right. The only way to get ahead if you go into a man's world *(he didn't think I would make a good teacher or nurse)* would be to work through men. The best way to do this was to find a boss with a daughter like me who would help pave the way. It took me some time but I finally did and I lucked out because he was also a lawyer. I noticed the picture of his daughter on his desk. He would be the one who gave me the confidence to test the system.

THOSE EARLY GE NEWS YEARS

My early years on the GE News in the 60s were frustrating because I worked very hard to get recognized as a business writer which was considered a man's job. As an intern on the Wheeling News-Register, I was sent on assignments as if I was a general reporter. Yet my first job on the GE News was to write a column called "Something for the Girls," cover activities at the GE Athletic Association, put together the list of service awards of those with 50, 45, 40, 30 and 25 years which often took half a page, get photos arranged of new retirees with just one line of information because there were so many and call the hospitals for a "get well" column or the funeral homes for the "sympathy" column as well as handling the free ad page for employees.

The "Something for the Girls" column bugged me the most because it was assumed they were only interested in recipes, fashions, and good housekeeping tips using, of course, GE appliances. While promoting electric heating for the home, I did manage to get into some technical writing and mention jobs growth in Schenectady as a result. I also looked for examples of women in so-called professional jobs and got in trouble with one I entitled, "The first Lady of the Lab." It was about Katherine Blodgett who had developed low reflectance "invisible" glass used for camera lenses. The problem was calling her a first lady.

The Civil Rights Act was being discussed in Congress and the company legal staff did not want it to look like GE was just now getting on board. So no more "firsts" I was told when describing jobs being done by women. I had already learned that during World War II, lots of women took on the jobs of foreman but when the men came back from war, it was again an all-male profession.

As for the compiling the sick and death lists for the GE News, it was just a matter of checking the spelling of names I got from the hospitals and funeral homes.

But I looked forward to meeting those long service employees, particularly the 50-yearones who had actually met Edison and Steinmetz. In those days, it was legal to hire teenagers and we had some 70 members of this group then. These contacts became very useful when Gene Folkman, who did a Pensioners newsletter, retired. It was decided to incorporate the Pensioner's Page in the GE News. I quickly volunteered for that job. I was to discover the meaning of the phrase, "on the shoulders of giants," when doing stories on early electrical pioneers. I was thrilled to interview Dr. Coolidge, who invented the x-ray tube in 1918, at his 100th birthday party. I also interviewed Dr. Ernst Alexanderson who came up with an alternator in 1922 to broadcast radio and later TV (*I have a photo of him with his 3-inch TV screen*).

But it was some of the unsung heroes who touched you the most. One was a radio pioneer, Rueben MacDonald, who had several patents involving radio transmission. In 1964, we found him at the age of 97 in a retirement home near Grant's Cottage below Glens Falls, New York. He had no peers left at that age and as an orphan with no family he was pleased to talk to me and the photographer. He was born in Oklahoma and did not know who his parents were. At the age of eight, he was run over by a stagecoach which resulted in his lower legs having to be amputated. A rancher took him in and as he put it, a lady visiting from the east was appalled to see that he was earning his keep by breaking horses. She brought him East, sent him to school and he ended up in Bldg. 36 at GE where the first TV broadcast was made to Proctor's Theater in Schenectady. While some speculate that today's versions of a hunchback Steinmetz or a near deaf Edison might have to use affirmative action today to get a job, I think of Reuben also having a tough time getting a job today.

So editing the Pensioner's Page for me was interesting and I did not mind the extra work. When the assistant editor left for another job, he was not replaced so I added columns of "Things in General" and "Around the Company" and started an "Out of the Past" listing since I did not have enough room on the Pensioners page. I also got the assistant's job of rewriting articles from department newsletters which gave me the chance to prove I could be a business writer. I would visit the factories in that department for more information which sometimes irritated the writers for those departments. However, Denniston would explain it was in the GE family's best interest to make sure everyone was aware of what was going on.

Denniston, older and wiser, was no longer interested in proving how adept he was. He was still getting accolades for his interviews of department heads on the state of their business. His favorite, however, was with Ronald Reagan when he came to the company's annual meeting in 1958 in Schenectady. While celebrating his 90th birthday now, he remembers how Reagan confined his remarks to just 30 minutes because that was how long the contacts produced then could be worn without using drops.

So when I volunteered to do the interviews with general managers, he was fine about it. He was into editorial writing and likened it to Ronald Reagan's support of capitalism and free enterprise. As a skier, he wrote a column, Sitzmarkers and Schussboomers and often went to lunch with the writers of columns called Kegler's Corner and Down the Fairway.

It was great fun working with him although I was pretty mad at him once when he sent me out on assignment and ate my packed lunch while I was gone. Each Friday after the GE News was delivered, we would set down and decide what news items we would pursue. We would also come up with a question I would later go out with the photographer to ask employees about. They were worded to get the best possible answer such as "What can we do now to make the company more competitive?" or "What contributions can I make in coming years to benefit our company? I got pretty good at spotting employees who would answer these with enthusiasm. But as the photographers Frank Warner and later Lou DiCerbo, Mike DeCata, John Papp and Frank Cusano would say, to get the answers I wanted I would frequently pick someone shy and ask them if they agreed with my thoughts.

There were times when I learned not to have an opinion. One embarrassing moment came when I was interviewing an engineer who had received a patent on a device which would transfer the load from one motor to another if it was going to overheat. Not realizing the motors were industrial and often in redundant rows for power, I suggested I did not want a second motor in my car. It didn't help much when I told him I was joking and that maybe not having that technology would be good for me because I was already at the limit in speeding tickets. He gave me one of those "Something for the Girls" cheesecake smiles and handed me a brochure describing the Small AC Motor and Generator business and its industrial, military, agriculture and consumer markets.

So I approached engineers after that with caution, even when I read everything I could find out about their particular technology. It was a lot easier with the general manager of the product departments. I had taken a lot of economic courses in college.

Back then it was all about sales and jobs. I remember each year we would publish a companywide pie chart showing where our sales dollar went. It was always in the neighborhood of 46 cents for materials and supplies, 6 cents for taxes, 5 cents for profits and 43 cents for employees.

Each week we would have a headline such as a picture of new equipment with the headline of "Investments Like These Improve Productivity, Create Jobs" or a story about improved products being introduced such as "Demand for Large DC Custom 8000 Motors Results in 20 New Orders" or the typical announcement in the mid-60s by the Works Manager, A.C. Stevens, noting that "we expect to be able to get enough new business to keep employment levels at around 25,000 in the main plant."

There were another 7000 working at the nearby Knolls Atomic Power plant, the West Milton Test Site, the Malta Test Station, and the Research and Development Lab. There was so much business travel that an airport limousine service made a visit to the plant every morning.

In those days, there were three shifts of factory workers maximizing the use of equipment. The parking lots were also always full including those at the GE Athletic Association. The GEAA offered twelve bowling alleys, a baseball field, tennis court, a bocce area (*many immigrant employees were Italian*) plus an archery and dart board area.

It was covering the bowling tournaments at the GEAA that I soon learned what was acceptable to be published and not published. There were over 100 teams from different product groups throughout the plant competing, including retired employees. One of these teams had a greater handicap than all the others because they were made up of employees who had been injured on the job. Some had the visible disability of the loss of a leg or an arm. I was so impressed at their esprit de corps and vigor that I put together a story with photos of that team. Needless to say,

I was told to get rid of it when the proofs came back for approval because to others it looked like GE was mangling its workers.

Safety, of course, was something we were always pushing in the GE News *(obviously both the company and employee benefited from safe work practices)*. I had learned from Day One never to let a factory employee not wear his safety glasses when a picture was being taken. But one got by me and I noticed it on the camera ready proofs at the printing company. It was too late to drop the photo and rearrange the page so I just took a pen and drew in some safety glasses for him. When the paper came out, despite my best efforts, you could tell the glasses had been added. However, the laughter at this did not result in a bad job review. Fortunately the safety manager for that department said it was great because all the employees now knew what they would look like if they did not wear their safety glasses.

The popularity of the adlet page, in which employees could run free personal ads, was undeniable even though Denniston and I liked to think it was our news reporting. The adlet page was also time consuming because ads were mailed in and had to be opened, collected and typed under labels of "for sale, wanted, wanted to buy, for rent, wanted to rent" and "free." I resented the fact I was still supposed to handle all of this despite having absorbed the job of assistant editor.

I learned the hard way to check the phone numbers for ads like "free manure." Invariably it was a disgruntled employee using the phone number of his boss. Others would get more creative in the "wanted to buy" ads. There was the false ad seeking a condom, not a condo, under "wanted to rent" hoping it would get printed. But it was mostly the "wanted" ads, that ranged from a case of beer to a night on the town and the "free" ads for various types of manure as well as for mothers-in-law which I quickly deleted.

One got by me that ended up in Playboy magazine. It brought huge laughs from employees *(glad we didn't have e-mails then)*. It was a "for sale" ad sent in by a retired employee and he actually did have the items for sale. The items for sale were a comfortable bed and upright organ. The fact the ad was real did not prevent me from being chewed out for not catching it. The editor of Playboy didn't help by noting that the assistant editor of the GE news in charge of the adlets was a woman.

However, it was after this they allowed me to farm out the typing of the adlets to the Typing Bureau. I grimaced when I saw the bill for that each week was close to what I was making each week.

Fortunately we had a Works Manager, A.C. Stevens, who was three bosses up and ultimately responsible for the GE News. He had a sense of humor and was understanding. He was already aware of my struggles as a woman when he discovered that I was having a problem with the head of the YWCA where I was living. (*As noted, a single woman coming into town was looked upon as having nefarious reasons for renting a room.*) Anyhow, the YWCA director, a Mrs. Fye, went a huge step further than they had in the Wheeling YWCA where there were no curfews. She had a weekday curfew of 9 p.m. and midnight on weekends. On Thursday nights, when I was putting the GE News to bed after legal and other changes, I was often past curfew and punished. I had been moved to the basement room where the closet was barely usable because of the sewage pipe that went up through it, which along with the furnace next door, was noisy and smelly.

One Thursday night when it was past midnight I could not raise anybody to let me in the locked door. So I went back to the plant and slept in Stevens' office which had a couch. That morning Stevens happened to come in early and found me there. The end result was him calling Mrs. Fye and telling her that the GE Capital Fund Drive that GE was running for the YW was in jeopardy. When I returned that evening, my new room overlooked the entrance and I had a key to the building.

Stevens was the kind of manager who even participated in a back page feature we had where we put short announcements of the many GE organizations such as the Foreman's Association, the Turbine Supervisors, the Apprentice Alumni Association or the Cosmo Club *(for singles)* and GEAA events. Most popular on this page were the photos of employees who raised the biggest squash, tomatoes, zucchinis and pumpkins, you name it. In fact, one fall Stevens' pumpkins came in second.

Stevens was so employee-minded he answered questions from employees once a month in a GE News column titled, "The Grapevine." While most of it dealt with questions from union members about grievances or suggestions for improving working conditions, he never hesitated to seriously answer a question. I remember the one from a worker who complained a couple of his co-workers, who were bowling in a GEAA tournament, were holding up the assembly line at work because they were just using their left hand to make sure their right hand would be in good condition for the tournament. Stevens replied they have a bum supervisor if he lets the fellows get away with this. Hobbies are o.k. he said but should never take precedence over our jobs.

You could always find Stevens at employee events like the annual Quarter Century Club outing which attracted some 10,000 employees and retirees with over 25 years of service each year. In his reports, A.C. always gave a yearly report which often started out with the need to attract new business to keep employment levels at about 25,000.

Stevens was a company man in the best sense of the word. He looked upon the GE News as a vehicle to promote job satisfaction and a sense of shared responsibility by all for all. It was understood it was a company house organ and he would point out to the unions they had a pretty good one themselves. He golfed with some of the union heads. He was never remote, always rooting for a GE team concept.

He understood an incentive was needed for factory workers paid the same daily rate and urged us to promote the Suggestion Plan to not just reward factory workers for suggesting ways to improve quality and save costs -- but also to give them the recognition. It also helped us fill the GE news each week. Because there were so many receiving suggestion awards, a winner only got their picture taken with a write up if their award was $100 or over. We were publishing about five of these each week. These interviews reviewing how they had improved business resulted in me learning a lot about the various jobs in the plant. However, because we did not want to upset the union heads, we would be careful not to mention the idea might reduce labor time.

Stevens was certainly not an old fogey either. Today I cringe about my willingness to adorn the GE News front page each week with photos of good-looking women serving as cheesecake. Summer issues were very popular because the models could be scantily dressed, particularly if it was a beach or pool scene. Every holiday, every benefit reminder and even most of the appliances one could buy in the Employee store, were "advertised" with a good looking model. This assignment also resulted in me becoming very popular in the GE Cosmos Singles Club and Schenectady Wintersports Club where I found many of my posers. One of my favorites, even though it shut down production in the main bay of Bldg. 273 for a while, was having a skiing friend of mine, Elaine Egnor, sit in the big 50-ton hook of the 500 ton crane in Bldg. 273 as if she was going uphill on a J-Bar lift.

While I did try to convince my bosses that a beefcake picture might interest our women readers, they were not interested. Still, I was able to sneak Fred Thompson in during a Valentine's Day photo which got rave reviews.

Actually there was quite a shortage of women in the plant. Most were secretaries although there were about 35 key punch operators (*computer operators today*), and a couple of engineering technicians (*some had engineering degrees*). But there were lots of young men on the Manufacturing Training Program, the Creative Engineering Program, the Advertising and Sale Promotion Program and the Marketing and Finance Programs....well over 2000. I got hold of a couple of the mailing lists and decided to invite them to a party my roommates and I were having (*I had finally saved enough money to move out of the YWCA*). It turned out that some 75 guys showed up and left early because we were the only three women there.

Still, I remember this party best because one of them sold me a car for $50. It did not have a floorboard on the right side and six months later it took 12 quarts of oil to get it to West Virginia to visit my parents. But as John Munter, now a building contractor in Saratoga, reminds me, he had nothing to do with the dent in the side when it got hit by the GE train (*I always seemed to be in a rush to get my job done*). The car, stripped of parts by my father, is still rusting away in a field ditch. My father, after convincing the Bank of Romney in West Virginia that I could pay for a 375 horsepower Ford he had his eye on, sent me back in it after we tested it.

Now I had another bill to pay in addition to the college bill I owed to my sister. While Denniston was all for giving me a raise, his boss Hal Reed pointed out that I was doing pretty well on overtime which I got because I was punching a clock. Even though so-called professional employees were not supposed to punch a clock, I had been told all the secretaries would be mad if I did not have to stand in line with them. And as time went by and I absorbed more jobs, the overtime did add up.

Denniston was not faring much better than I was under Reed's reign as the plantwide communication manager. Reed was one of those managers who felt employees should be so grateful working for him that they would want to help him at his home. Denniston was sometimes called on to tend to his horses on weekends when he was away with his family. I was told I could enjoy Saratoga Lake while I was babysitting his young son who he felt was not old enough to go on the boat with them. I soon found out the problem was not that he was too young. He was a spoiled brat. Worse, when he stomped a cat to death in front of me, I decided he needed a lesson and gave him a good pounding. Again, I considered myself as having lucked out because Reed decided the secretary in community relations could do a better job.

Today, I cannot believe how timid I was about not asking for more money and buying into the line that I should be grateful I had a man's job. But I was having a good time and I could get more money by working overtime. In those days, guys also paid for the date. My only problem on $5200 a year boosted with overtime was budgeting enough money for food and rent and to buy clothes and shoes in addition to now paying the car bill and what I still owed my sister.

To this day I regret spending a week's salary on a mink collar to go to the University Club in Schenectady's Stockade where all the Ivy League trainees lived. All I attracted was an accountant from Connecticut who was a real bore and asked me where the rest of my coat was. I soon learned to spend any spare money on ski equipment and get on the bus which was filled with manufacturing and engineering trainees.

Still, I was having a lot of fun and I was enjoying being part of the business world learning how sales and jobs were created. However, in 1961 I came close to going back to Wheeling and not by choice. A nationwide strike was called by the International Union of Electrical Workers. By this time, I had become familiar with knowing who was big in management. One of my jobs was delivering the proofs of Friday's GE News to the plant lawyer, Emil Peters, for his approval. His office was near the area Vice President of Power Generation Bill Ginn, who was known for his flamboyant style and disregard for plant rules. Some wondered why he bothered working at GE since his wife was heir to the Colt fortune. He always seemed to be amused by everything.

I sure found this out one day when I strayed too close to his office entrance to see Mr. Peters. I had noticed a bronze grill in the floor at the entrance to his office but this time I stepped on it to avoid a delivery man. All of a sudden I felt bristles coming through my sandals. In my confusion, I yelled, thinking it was something alive. The delivery man was laughing and calming me down when Ginn came out of his office. He motioned the delivery man into his office and then amused himself demonstrating his automatic shoe brushing machine to me.

He picked up on my West Virginia accent in the process and from then on I was known as Miss West Virginia. He asked me why I was not wearing heels like the rest of the women. He had an even a bigger laugh when I told him Hal Reed, my boss's boss, was short and did not like me towering over him so he had said wear some nice flats. Because I had a size 11 foot, I had to order them out of North Carolina and the pair had not arrived yet so I had on my summer sandals. I told him I was not complaining... that I was glad not to have to wear heels because they were not only expensive but it was difficult getting around the factories in them, particularly in the foundries which had dirt floors. This resulted in more laughter from him and an invitation into his office. When Ginn flipped a switch on his desk, a portion of the mahogany wall slid open to reveal a bar. The delivery man began restocking his supply. I was stunned and then told by Ginn that he assumed Miss West Virginia could keep her mouth shut to security. I assured him I could and solidified that promise by telling him that he shouldn't worry about the guards, that they often shared the beer with me they took off the night shift employees trying to sneak it through the subway gate. He just laughed and told me to keep up the good work on the GE News.

So when the nationwide strike spread to Schenectady and everyone was locked out of the plant by picket lines, our communications group was busier than ever. All of us, including Bill Ginn, moved into the swank, elegant old Hotel Van Curler where I assumed he had the top floor which had a presidential suite. But I was to see him again. We were still able to publish the GE news outside the plant and find ways to have a bundle accidentally drop off a truck going by the plant entrances where thousands of electrical workers were circling to keep everyone out until their demands were met. We knew the GE News would be read because everyone liked the free adlet page. The city newspaper, losing this business, had complained about it but to no avail. It was considered an employee benefit.

One of my jobs was to deliver news releases which were practically written by lawyers instead of our editor and the press relations staff. The National Labor Relations Board was checking every word to assure themselves the company was not violating any policies to get people back to work. I also read instant news announcements to local radio stations. One of these was WSNY where I had developed a liking for Steve Fitz, a newsman there. He did not take the labor-management situation as serious as many did. He was friendly and very popular in the community for being fair-minded. The pressure was on to support GE but he gave both the union and GE an equal amount of time to comment on the strike. He would take calls from both sides and replay them in an hour-long program. He knew that both the union and GE would be providing false "callers" to promote their position.

I had become known as the bowler's wife who wanted her husband to go back to work instead of bowling all day, noting things like the children were getting hungry and the company was making a fair offer. The union side answered with a bowler's wife who said he should be on the picket line instead of bowling, that there was not enough money for food and clothing for the kids. Fitz used to tease me about being such a good company girl and wondered how many factory workers had wives with southern accents.

But, unbeknown to me at the time, Fitz also had a live evening program where people called in on a certain topic. On that evening, it involved a shopping trip the Schenectady Woman's Club was going to take to New York City. That was also where national negotiations were being held involving some 150 U.S. GE plants. When I called in, I did not realize it was a live show and Steve answered asking me what I was going to do in New York. I immediately replied that Bill Ginn and I were going down there to settle the strike. It went live before Fitz could cut me off.

Before he could get back to me to tell me what happened, the door opened and in came my boss's boss, Reed. As bad luck would also have it, our room had a TV set. I was watching Dick Clark's bandstand show when he rushed in. He turned it off before I could get to it and fired me. Denniston came in and tried to save my job. But two more layers of bosses arrived all agreeing that it was best I leave.

But fortunately for me Ginn's wife had planned on going on the shopping trip and asked her husband about it. By this time I had my coat on and was about to leave when the door opened and in came Ginn. I tried to apologize. But he just stood there with a big grin as others told him they had handled the situation and that I would be leaving the job. He just laughed and said: "Get back to work Miss West Virginia and try to keep your mouth shut. Don't tell the whole world we're going to New York."

Seeing the shocked looks, he told them somebody should show Miss West Virginia a good time in New York. And, looking at my short boss Reed, he added, "Tell her you have to wear heels down there."

A year later, when Ginn was tried for price fixing steam turbine prices with Westinghouse, no one was sadder than I was. Agents were all over the place and scanned every edition of the GE News to see if we were using it to let Westinghouse know what we were bidding. Even the head of union relations, Red Levy, who had become a close friend to Ginn, was upset over Ginn taking the fall. I had heard the whispers that Levy would never be promoted to the top corporate union relations post because he was Jewish. I liked him because he and his secretary, Judy Love, taught me how to play bridge at noon hour. He treated everyone with respect, no matter what their position in the company was. Unlike one of the sanctimonious department plant relations managers, Carl Hudson, who I had heard call the strikers "animals," Ginn regarded the union factory workers as very competent adversaries.

I will never forget the noon hour that a call came in to Levy from Ginn when he was in prison in Philadelphia and the press was reporting that a guy doing life for murder was in a cell next to him. Levy asked him how things were going. Ginn joked that he guessed Levy had heard that they were now accusing him now of rigging the electric chair. While the charges stuck, I felt that a higher up had to know what was going on. But it didn't surprise me that Ginn would plead guilty. It was another experience for him and he was in the end, a company man to take the fall.

His replacement, Donald Craig, was a gentleman and a nice guy but not nearly as interesting as Ginn. I would find myself outside Craig's house one night wondering if I would get fired for trying to keep myself from being fired. But that comes next. If I thought I was working long hours then, I soon got a real test for my stamina when the company decided Schenectady needed to be more competitive. A major campaign was launched in late 1963 which drew nationwide attention because labor unions were perceived then as having too much control.

MAKING SCHENECTADY COMPETITIVE

Schenectady GE became major U.S. news in April 1964 when it launched its Make Schenectady Campaign (MSC) involving one of the largest unions in the nation, the International Union of Electrical Workers (IUE). At the time it was felt unions had too much power and that the country's National Labor Relations Board (NLRB) was too soft on union power.

Jim Carey, who headed the IUE during the nationwide strike against GE in 1960 involving 70.000 members railed against GE's "Take It or Leave It" approach. The NLRB agreed. Back then GE protested that it had willingly entered collective bargaining, that it had not sought to get rid of the union, it was not bargaining directly with employees and that its communication to employees was legal and in good faith. The unresolved 1960 case was still under appeal by GE in 1964. So when the company decided to correct what it felt were pay inequities according to job skills, some 150 GE plants as well as other industries were closely watching as well as the NLRB.

"A Fair Day's Pay for a Fair Day's Work" was the official mantra in the Make Schenectady Competitive program to get rid of piecework in which a worker was paid for his individual output. Day workers were paid according to their rating in the job evaluation system which had some 30 classifications. Plant management argued that not only should jobs be evaluated more closely to skill but that piecework pay had gotten out of control, making the plant non-competitive with competitors.

Basically, the deal offered to the Schenectady unions was signing a deal in which GE would invest $60 million dollars in new equipment and buildings and not move out the motor business if the union would agree to a Fair Day's Pay for a Fair Day's work by eliminating piecework and agreeing to a job reevaluation program.

Unless these actions were taken, it was announced the plant would lose 5000 jobs. If the agreement was reached, there would be as many as 5000 new jobs created because new investments to modernize the plant and the agreement to a "Fair Day's Work for a Fair Day's Pay" would make the plant competitive. The investments included a new $20 million Large Generator and Motor building as well as modern equipment throughout the plant to make it more competitive.

The new Schenectady GE vice president, Donald Craig, said there were gross inequities between the pay of over 3000 pieceworkers and the 6000 day workers. He cited compensation reports showing nearly 1000 of the plant's 3000 pieceworkers made over $10,000 a year which was higher than the salary of some engineers.

Negotiations were even more intense between the local IUE and Schenectady management because the outcome could affect the bargaining of other national unions and companies. So it wasn't just the NLRB watching GE in Schenectady. Labor across the country was concerned.

As soon as the MSC offer was explained, GE was quick to hire independent pollsters such as Opinion Research in Princeton, New Jersey where the poll showed 12 to 1 in favor of the deal.

Such a program with a national impact required a massive communication effort. Denniston left the GE News to join others in press and public relations to put together information booklets as well as write speeches for management and news releases before and after each meeting with the IUE. Among these were Jack Callahan, Fred Haas, Bill Folsom, Dick Healey, Bill Kennedy and Dee Logan.

My job would be to continue writing general interest GE News articles as well as go out with the GE photographer to ask specific questions to support the MSC program…questions which were often leading ones such as "Would you like to see 5000 new jobs if the union agrees to a fair day's pay for a fair day's work?" or "What can you do in your job to be more competitive?" Naturally these were sensitive questions in the shops and I really had to be careful to whom I posed it.

Now alone in the GE News office, I got my first taste of spending even longer Thursday nights at the printers waiting for approvals to get the weekly paper out on Friday. Normally, we needed only a couple of approvals but now several lawyers were involved because the National Labor Relations Board, which the company felt was full of liberal Cornell lawyers, was watching what was said and done.

Every word in negotiation articles was checked and rechecked. While Journalism's Fourth Estate rules involve attributing a quote to the source, having our Vice President Craig comment on job skills was a no-no with the NLRB. They had determined that job skills could only be determined by an agreement between the company and the union. Thus quoting Craig would not do. After that it was, the company said this, the company said that or the union said this or that – but never attribute a quote to an individual during negotiations.

Making it more difficult was a new set of proofs had to be sent by cab to a couple of lawyers as well as added management to show every change had been made, no matter how small it was. During the six months of negotiations before the union finally agreed to the MSC program it was rare to get out of work before midnight on Thursdays.

While whole articles would often be yanked from the proofs, even little changes like adding one or two words kept me, a bank of linotype operators and the composing room foreman wondering if everyone had lost their mind. Actually it was easier to replace a killed article because I always could replace it with a story touting GE benefits which was already set in type for such occasions. I also had a backup page of GE pensioners relaxing at their retirement camps and some of them spending the winter in Florida.

Photographs were engraved on metal plates for reprinting and fortunately could not be changed-- just deleted if offensive to the NLRB. However, changing the text was a nightmare. A bank of linotype operators would have to type in each individual letter in their machines which would cast it in hot lead. Once the letters were assembled, the composer would then take the lines of composed type – set upside down and backward – and secure it in the metal framed page rack.

It would take at least 15 minutes to make a change and then more to run a proof over the inked metal type to make sure the change had been put in place. It usually meant a whole paragraph would have to be reset to make a change that overran the space if something could not be cut. It was tedious work and work that I could be easily fired for if a correction was not made.

GE had gone all out to convince the union they must agree to what was constantly billed as a "Fair Day's Work for a Fair

Day's Pay." Press kits for local and state government officials were put together and radio and TV time was purchased. In addition to hiring more press and community relations personnel, a new government relations manager position was set up to deal with state officials and politicians. It was headed by Hal Halvorsen.

The prestigious Hotel Van Curler was still operating so it became a place for Community Relations to hold a hospitality hour for select government and business leaders in the area on Mondays. The Van Dyck, in the historical Stockade area, was chosen for other important community leaders including small business leaders and heads of not-for-profit groups who had a lot of influence as well as church leaders.

But the most popular place was Jimmy's bar on Albany Street which hosted MSC talks to less important civic and business leaders. It was also the favorite watering hole for the press in the area as well as for those of us from GE pitching the benefits of MSC. It is a wonder all of us, particularly those of us who went to all three information parties, didn't become alcoholics.

We often had drinking contests at Jimmie's where there was also a piano player who would serenade the winner. Al Barton, the community relations manager, Fred Haas from press relations, Buddy Ottaviano from the Gazette and Denniston and I often won these. One particularly raucous evening involved a last-two-standing contest between myself and the community relations manager. Denniston decided to aid me by telling the bartender not to put any brandy in my Brandy Alexanders (*I was still into exotic drinks*). Nine of these later had me as the winner but I was feeling sick with all that cream which I threw up. By this time Al was at the piano getting ready for a serenade when he spilled his drink on the piano player. That was too much, even for Jimmy, who ran a light ship. He told us not to come back. But his business side did prevail and we were back in just three weeks.

It is a wonder that Barton did not become an alcoholic running all of these parties which he fully participated in, particularly the cocktail part. On one occasion Jimmie told me to drive him home which I had trouble doing because Barton insisted he could drive his new Plymouth which had push buttons for the starter as well as the windows. To start the car, he kept pushing the window buttons which eventually trapped his lopped-over head when going up so it was then I took the wheel. He wanted to go to the Edison Club to pick up his MSC briefcase which I discovered also contained a bottle of Baileys. When he put some of that in the coffee I had ordered for him, the Edison Club bartender, not appreciating Barton bringing his own bottle, told us it was time for us to go home.

On the way out, Barton decided to go to the men's golfing facilities. When he didn't come out after a while I went in to find him fully dressed and soaked from being in the shower where he went to sober up. Still, I did not let him drive. When we got to his house, he insisted he could make it into the house which was framed with shrubbery. I told him I would pick him up for work the next morning and watched as he went up the walk. When he was at the door, I left. Several hours later I got a call from his wife. She had called Jimmy's to find out that I had left Jimmy's with him and demanded to know where he was. I told her I had dropped him off and I had the car because he had not been feeling well and that I was to pick him up for work the next morning. She decided to call the police because I had dropped him off hours ago. When they arrived at her door, they noticed Barton lying in the shrubbery beside the door.

Looking back, I can't believe how much we drank at these MSC events to rally support for a Fair Day's Work for a Fair Day's Pay. One time Barton hosted an event at the Ten Eyck in Albany for state legislators. He and couple of the other press relations guys picked me up at Fox and Murphy's ski store where I bought a pair of boots. On the way over, one of them tossed one of my boots out of the car when he was looking for his MSC kit. Barton, always the gentleman, stopped the car and retrieved it. So after a couple of drinks it was Barton I looked to for help when I went to the john and discovered I needed a dime to get in the booth. Barton could not believe that such a fancy hotel would have pay booths and told them he would never host a party there again. He went to the ladies room with me. He did not have a dime either so he assisted me over the top. It was then the Ten Eyck manager informed Barton he hoped he meant it when he said he never wanted to host a party there.

I was glad none of these Al Barton community relations "events" were on Thursday night because that was when the GE News had to be put to bed. Looking back I am also glad I was as healthy as I was. I was getting about five hours of sleep a night from all the long days at and after work which led up to a pressure cooker on Thursday nights at the printing plant. We all had a laugh when we were celebrating my five years of service at Jimmy's when it was mentioned that I was now eligible for an award given then for never having taken a day off sick. It was then that Reed mentioned it would not be fair to the other women in the plant because I had had a hysterectomy and therefore did not have to deal with the problems they did. I had completely forgotten about putting this on my job application as a way to get hired. This caused Barton to order a round of Bloody Mary's for everyone.

At the height of the six-month long program to get the union to accept "The Fair Day's Work for a Fair Day's Pay" deal was a special edition of the GE News devoted entirely to MSC in a format we had used for national negotiations. Everything was laid out in a four page spread showing all the benefits that would result -- photos of new buildings and equipment which would be made if the union would agree to give up piecework and a new job evaluation program. I was there waiting for final approvals of proofs which had been sent out by cab time and time again as I made changes and substituted articles in this special Make Schenectady Competitive edition of the GE News.

On the front page was a photo of Craig who as the head of power generation business had the most to win in the agreement because they had the greatest number of pieceworkers. Unknown to the methods specialist, one of them had made a device which he attached to his machine which enabled him to triple his output. He was making more money than most of the engineers and no one could figure out why because he took the device home with him. Others, through sheer energy could nearly double their pay.

Also on the front page was the photo of another GE Vice President, Oscar Dunn, who headed the Motor and Generator business which had a lot more day workers because it was more of an assembly line business. Dunn had not wanted to go along with the agreement even though it would include a 200,000 square foot addition to his business. He resented the fact his experienced factory employees would often leave for the better paying steam turbine jobs. He had really wanted to move the entire motor business to Fort Wayne, Indiana where the community rate for labor was cheaper. But Craig, as head of the largest and most profitable business, had area responsibility.

The linotype operators had gone through the usual horrendous number of changes as lawyers surveyed each word, adding or deleting words here and there to assure themselves they were not in violation of NLRB rules. During this six-month campaign I had gotten used to spending Thursday evenings at the printers just making changes in a page or two of MSC News. But this special edition had everybody on edge because of the sensitivity of laying out the whole final deal on four pages for the union to accept and to make this final offer within NLRB guidelines.

It was past midnight and I had been informed that if we did not have the approval by three a.m., it would be impossible to have 30,000 copies of the GE news printed and delivered in the plant on Friday morning. Another 2500 needed to be in the mail for Saturday delivery to community leaders as well as the normal 10,000 retirees. Despite my constantly reminding those approving the GE News as I sent cab after cab to their homes to get their approvals and then resent to show them changes had been made, each was making changes on each other's change. Finally, at 2:30 a.m. I got the final approvals.

A cheer rang out as several linotype operators, the pressroom foreman and I started over to the nearby Tip Toe Inn to have a few drinks celebrating the overtime pay we would be getting. I had gotten used to and began enjoying the toast they made to me to "keep up the bad job" because they liked the overtime. However, they knew from all the cab drivers that it was not me making the changes. They knew full well I would have been fired by then if I was that incompetent.

Just as we were about to leave, I got a call directly from the Vice President of the Motor businesses - Oscar Dunn. I knew it was him because I had covered different events in his division, mostly when he was giving out awards to try to motivate his day workers who received the same pay. He had gotten my number from his division newsletter editor and his message was simple. "Do not print that edition. I did not approve of it."

I was speechless. All I could do was just thank him and hang up. I immediately went into action -- first it was a call to Denniston, then his boss, and then his boss and then the PR and Government Relations Manager and the lead lawyer who got proofs. None of them seemed to be home. I tried the bars at the Hotel Van Curler and Van Dyck but they were not there celebrating the final MSC offer. It was closing in on three a.m. If I did not get an o.k. by then it would mean there was not enough press time to get the GE News into the plant for distribution as usual.

Now the pressure was really on. It was a Catch 22, I said to myself. "Be fired if you do print it or if you don't print it." I saw only one choice to keep both from happening. I knew that Craig had area responsibility and that if I got his o.k. I would be safe.

I roared off in my TR 3 sports car (*overtime had allowed me to buy a used one*) to Craig's house telling the crew to go ahead to the Tip Toe Inn but to stay there until I got back in case there were changes. Craig's address was in the phone book and even though it was an elegant neighborhood, it was not that far away. I could get there in ten minutes. When I did arrive, the house was dark. But there was a car out front with two people in it. Even though they looked like they would rather not be disturbed, I knocked on the window. The girl turned out to be Craig's daughter. She could tell by the sound of my voice that it was imperative her father be gotten up which she did not want to do because she was past her curfew. She had planned on sneaking in the house. Since I said I was going in with her and she was saving me from being fired, she agreed and said goodbye to her boyfriend.

Craig, as sleepy as he was, was a perfect gentleman when I explained the phone call from Oscar Dunn. He just smiled and signed the proof I handed him saying it was o.k. to go to press.

I got back to the printing plant a minute or so before three a.m. to give the "go-ahead" again for printing and joined the gang at the Tip Toe Inn. Bill Mercer, the composing room foreman, had not one but two Rob Roy's waiting for me containing his favorite Scotch. We celebrated our amazing effort to get it out in time to be distributed to the first shift coming in the plant.

I also celebrated not having to go back in the plant to punch out the workday on the time clock at three a.m. I was still not considered a salaried professional employee in the mid-60s. I had to punch the clock like the secretaries, low-level office employees and factory workers.

The linotype operators, tired of setting all that type about being more competitive on the job, suggested that they were going to make me more competitive by helping me save time. They gave me a rubber stamp with type on it that matched the type on the time card machines. I could just whirl it to the time I was finished and would not have to drive back into the plant in the middle of the night to punch in my time on the job. So we always had a ceremony of stamping my time card at the Tip Toe Inn. Looking back, I realize now that I should have built in the 15 minutes it would have taken me to drive to the plant for the official time stamp.

However, at the end of that year Reed realized that my overtime *(which was given to all employees who punched a clock)* had nearly doubled my wage. So he took me off the clock and gave me a level raise resulting in my making less money the following year.

It was the Make Schenectady Competitive campaign which resulted in me getting my name listed as editor even though I had been the only one on it for nearly a year without a formal promotion. I had been told that blue collar workers, particularly during these tense negotiations would not read the paper if the editor was a woman. After being on the MSC team, Denniston had joined the Advertising and Sales Promotion Department as a speechwriter.

Being well acquainted with the company lawyers, I knew there was one who had a daughter who was a senior in high school and interested in Journalism. His name was Emil Peters. He had her pictures all over his desk and I could tell he was quite proud of her. I remembered my father's advice that in working through men, find one that has a daughter. They would be the most sympathetic and supportive. So when I was running proofs by him, I inserted an old copy of the GE News with it that had a nameplate listing myself as a communications specialist and Al as editor. He spotted it immediately and asked me why the nameplate was no longer being printed. I told him that hopefully it was temporary because as soon as the Make Schenectady Competitive campaign was over, maybe my bosses would lose their fear of nobody reading it because it had my name as a woman as editor on it. He was upset that he had not noticed it missing these past six months and said that the nameplate must go on the paper because there were over 10,000 retirees getting it in the U.S. mail. He said it could not be in the U.S. mail unless an identifying nameplate was published in it. I pointed out there were also many local and state officials on the mailing list as well as community and business leaders and even a guy from Saudi Arabia who had written me when I put him on the mailing list that "Allah thanks you."

He told me to run the nameplate with my name on it as editor immediately. He added that when it came out the next day, if I had a problem with my management then I should have them call him. He even told me to be sure to use my full name as was done in the past because they might try to have me use y initials or even their own name as managing editor or something of the sort. He took the old GE News nameplate and put in my name. Now you have my final approval, he said. So when I went back to the printing company, I took out a filler benefit story and inserted the nameplate.

I spent nearly a sleepless night wondering what would happen Friday morning. The first papers were delivered to the plant works manager including my bosses. I was soon called in to explain the appearance of the nameplate. All I could say was "talk to Mr. Peters. He will explain everything." Peters had been right. They came back to me suggesting I use my initials in future editions. But I held firm that I only would do what the lawyer said was legal.

The calamity they thought would happen, of course, did not occur. A few phone calls to department employee relations managers, most of who were surprised about the whole issue, reported that employees were not tossing the paper out. One of them even ventured to say that as long as the free adlets were there, my boss could be assured the paper would continue to be taken home.

Thanks to Mr. Peters, I started to realize I needed to be more assertive. I was also secretly thanking all those women out there who were now burning their bras in protest for equal pay and opportunity. Negotiations had taught me that radical actions result in a new bargaining center and those bra burning women were making working women seem more reasonable just asking for small advances in pay and opportunity.

With this newfound confidence, I told Reed I wanted to make myself more competitive in Schenectady by having an official in-plant driving pass. There were nearly 300 buildings spread out over 500 acres and I could save a lot of time not waiting for the plant bus, hitching a ride or waiting for a photographer to show up. I was told I was still not high enough in the pay level to get an in-plant driving pass, to use the bus and quit hailing down battery truck operators because that was not safe.. He even added he had often been amazed how quickly I sometimes came back in from covering an event.

What I didn't tell him was that I had sometimes bribed the plant guards with enough GE pens and drinks on hot days that they let me drive in the interior of the plant instead of having to spend all that time walking to and from the big parking lots. But this could not go on forever because the head of security had already called a couple of them in for questioning when he saw my car in the plant. It didn't help that I was driving the used TR-3 sports car.

Having had great success with my time card stamper which I no longer needed, the linotype operators, after a round of drinks on me at the Tip Toe Inn, readily agreed to print a plant pass from a copy of one I had made of Hal Reed's identifying him as Manager of Plant Communications.

The pressmen put my name as Editor on it and all three shifts of guards celebrated the fact I finally had a pass and was now the Editor. The head of security knowing that Al Dennison had a pass when he was Editor did not question them. My only challenge was to park as far away as I could from Reed so that he would not know I had gained access to the plant.

However, my experience punching in at the time clock came in useful when Schenectady Wintersports Club President Paul Lozier and I were running an "anything goes" regatta on the Mohawk River which goes by the plant. It was a ski club but we decided to branch out with canoe racing in the summer. Our first-time venture was a huge hit. Heavy promotion to make people aware of the need to improve the quality of water resulted in some 300 contestants signing up. We had no money so awards became bottles of water taken from the river. It would be an annual event and we would tell the winners to save their bottles, hoping each year would show an improvement in water quality.

Renting time equipment was going to be a problem but the GE guards helped me out by donating a time card machine which we placed at the finish line. Cards were color-coded to designate each class and given to each entrant. After a mass start, timing would automatically be handled because all the competitor had to do after canoeing seven miles was run up to the river bank at the end and insert his or her card to log the time. Several problems surfaced. The cards were often wet as the anxious competitor jumped out of the boat for a fast start to the time card machine, jamming it. Worse, the machines only recorded by minute and often there would be six or seven competitors with the same stamped time because they had come in within that minute. Still, it was a fun day and the next year we had enough donation to rent timers that recorded seconds.

But back to Making Schenectady GE Competitive. While much of the front page of the GE News was taken up with Make Schenectady Competitive articles, there were 20,000 other salaried employees in the offices, labs and research centers to keep informed as well as gain their support for the program. So news as usual had to go on and I had to do all of that as the lone person and editor. I was often so harried to get everything done, I started prioritizing work and figuring out how to save time. I was still doing the pensioners newsletter but did not have time to make those joy rides to lakes to see how they were enjoying their life on pension. I used the Mailbag section to have them send me news and figured out a way to quickly cover their monthly meetings which featured a guest speaker.

Not having the time to sit there an hour to hear what was said, I would go early when the speaker arrived and have the photographer take a picture of the speaker with a couple of retirees I had lined up. Then I would take off and report on what I thought the speaker probably said. One time the speaker was late The association officers had gotten used to me lining up people *(mostly themselves or friends)* to run in the GE News. The speaker was the new City Manager Birbilis. None of us knew him. When a fellow in a business suit came walking near the Building 5 auditorium, I yelled "Mr. Birbilis" and a guy named Norm DiBelius said "Yes, that's me."

Before he could protest I had a photo taken of him with my readied group and was ushering him down the auditorium for his talk. As he was being announced on the stage, I noticed another man coming in the door and heard him ask for the president. I was able to quickly pull the confused and overwhelmed Norm Dibelius off the stage and replace him with the city manager. Yes, I had to stay through the whole talk to get the correct photo later to commemorate the event on the Pensioners Page.

Even after the MSC agreement was signed at the end of 1964, the union kept filing grievances over the part that involved the job re-evaluation section. Piecework had been eliminated and the transition to day work pay was taking place in incremental steps. But the job evaluations program was another whole can of worms with lots of articles about how more jobs were going up in pay scales rather than staying the same or down under a red circle program. The union never stopped protesting those that were red circled despite all the headlines and articles the following years showing how GE was spending the $60 million worth of investments in plant and equipment it had promised. Every week we ran a progress report on how the motor or turbine business was now more competitive and getting huge orders – all resulting in the 5000 jobs being saved as promised and 5000 more being added. The "A Fair Day's Pay for a Fair Day's Work," mantra never stopped.

To combat slowdowns resulting from an occasional job evaluation strike, reminders of the Fair Day's Work for a Fair Day's pay creed was supplemented in the GE news with the one line 36 point headers spreading across the bottom of several pages in large type reading: MSC Reminder Jim, Up and At Em with some Vim... MSC Reminder Flo, the Long Coffee Breaks Must Go....MSC Reminder Jake, Office Hours Start at Eight....MSC Reminder Fred, Keep Waste Down to Stay Ahead....MSC Reminder Jack, Make Sure Products Don't Come Back... MSC Reminder Ned, Give Your Best to Keep Ahead ...MSC Reminder Tod, an Efficient Plant Gets the Customer Nod...MSC Reminder Pete, Top guy, Little Guy, All Compete. Still we continued to run the original MSC mantra such as: MSC Reminder Kay, A Fair Day's Work for a Fair Day's Pay" or a variation of it such as "MSC reminder Joe, A Fair Day's Work for a Fair Day's Dough."

Finally, with full acceptance of the job evaluation program, the investments were made and there was much joy in the Schenectady plant as employment rose. Some new businesses were now growing and jobs added as a result of technology. One of these was the Power Tube department which had introduced the new Klystron tube to improve the accuracy and increase the range of missiles and satellite radar detection systems. In fact, it prompted Jack Callahan of Press Relations, whose spare time endeavor was with the Civic Playhouse, to rewrite the lines of "Seventy six-trombones led the huge parade" to "Seventy six turbines led the big parade with 110 transformers right behind. There were rows and rows of dynamos and tubes of every shape and kind." We often sang this at Jimmy's bar which all of us had grown quite fond of. Plus Jimmy preferred it over "MSC Reminder Jim, A Good Drink Goes to the Rim."

In addition to the articles hailing the success of the Make Schenectady Competitive campaign, employees were often reminded how even the New York Times had hailed the MSC agreement with the headline of "Statesmanship in Schenectady."

But as in most years prior to national negotiations, which this time came just two years after the local MSC agreement at the end of 1966, GE communicators were again called upon to do some major climate setting to resist demands from the union for major increases in wages and benefits. Competitors, particularly foreign ones, were described as voracious and bloodthirsty. Lost orders became the headlines of the day. If a new order was announced in the outside business press, we would attribute the success to new tools and equipment and the technology gains in improving the efficiency in turbines and motors. But mostly it was only reports about lost orders to lower employee expectations of wage and benefit increases. Most of the stories involved the motor division which had some 15 competitors in the U.S. including Westinghouse and many more offshore. Naturally every time a foreign competitor took an order that required a large headline.

However, the steam turbine business was still the biggest game in town and the union knew that there were very few competitors in this business because it required heavy investment in plant and equipment. And so did the management at the motor businesses. They wanted to move out to avoid strikes they felt they had nothing to do with. They also resented having their good factory employees using the negotiated bumping procedures for better paying jobs in the large steam turbine business. So explaining the company's need to become more competitive was never ending.

Even U.S. Secretary of Commerce Connors was brought in to discuss the nation's need to export more and be more competitive. Other state officials involved in commerce visited the plant to support the need for more exports to keep jobs in New York.

Bringing in U.S. and state officials concerned about exports did not bother the NLRB watchers. But my enthusiasm (*I now regret it*) for helping the cause nearly landed us in trouble with an unfair labor negotiations charge the company had gotten back in 1960 with its "Take It or Leave It" approach. A part of that edict involved unfair communication so when I started having our cheesecake volunteers dress up in foreign costumes to complain about the low wages their husband were getting. It was too much for the NLRB.

I was called in and told to stop it. The NLRB watchers reacted pretty fast to what they determined was unfair communication. There were only two that got published before the plug was pulled. One was a gal, dressed in a short Roman kilt, driving a horse-drawn Roman chariot *(I found one on a pony farm near Rotterdam Junction)*. She was quoted as saying you can't go far if you just make 65 cents an hour as a Nuovo Pignone electrical worker. Another was a Dutch girl who asked if you think wooden shoes are uncomfortable just try living on 84 cents an hour like Phillips Electrical Firm workers in the Netherlands do. Of course, it was noted that the average U.S. Electrical Workers make from $1.50 to $2 an hour more than their foreign counterparts.

I had gotten used to the thought of being fired by now and was not scared when called to the carpet for the photos. In fact, the new Works Manager Red Parker asked if I had any more…that even the union guys at the table had asked the cheesecake not be stopped because of the NLRB.

One of the biggest battles in the 1966-67 negotiations was the issue of collective bargaining. All of the diverse unions in GE plants, from drafting to plumbing to electrical, across the U.S. wanted GE to deal with them all at once instead of individually. This time they got a reprimand from the NLRB with their mantra of "Nobody Settles Until We All Settle" which sounded a lot like GE's negotiator, Boulware, in the 1950s who was known for saying "Take it or Leave It."

But pent up frustration over piecework pay losses and job evaluations -- plus national union pressure at other plants not wanting the same thing to happen to them -- resulted in a nationwide strike. Even an article that the Large Motor and Generator Department had bid on an order at a loss to assure they would still be able to be in business when workers returned had little effect. Massive layoffs at Westinghouse during strikes there were reported and rumors the motor divisions might move out were denied by GE. It could not look like a betrayal of the MSC agreement but such news and rumors would still have its effect in lowering employee expectations during negotiations. Even the struggling but growing gas turbine business *(which was not the money maker it is today)* got in the act by suggesting they might have to expand somewhere else.

But as soon as the strike was over, everything settled down and construction resumed on the projects that had been started under the MSC agreement. So my job became a lot easier and more fun now that I no longer had to search for stories of flesh-eating, voracious competitors or devise new charts to show the cost-price squeeze the plant was facing.

After a few months, we were back to happily reporting a profit of five cents on the dollar and the charts showing the slice of the pie for wages, materials and supplies and taxes. In just six months, the Large Motor and Generator headline was not taking a loss on the sale but it read, "Pay-Productivity Resolution Helps LG&M

Win Largest Contract in History" with a kicker of how the high efficiency of GE motors resulted in a win over a low bidder.

In those days, as soon as employees retired, new ones were hired and I often sought them out to tell why they wanted GE jobs. Headlines and stories of orders won and jobs created were back on the front page. I was running even more human interest stories about our GE family. Sure, there was the motive of improving productivity by creating job satisfaction and pride in workmanship. But it was also part of GE's paternalistic mode of operation that went back to creating the first industrial pension plan in 1912 and offering a GE Savings and Security Plan enabling employees to be stakeholders. Under S&SP, employees could buy GE stock and U.S. Savings Bonds which enhanced the feeling of GE being a family. However, it used to bother me when I got calls at year-end from GE accountants to run an article about unclaimed stock and U.S. bonds. In those days, stocks and bonds were mailed to the home of the owner who may not have changed an address or died. A surviving spouse might not even know the stocks and bonds existed. Today, federal regulations require both signatures on such plans to assure they are aware of it, particularly the Pension Plan which offers survivor benefits. I would look at these stacks of stocks and bonds which were unclaimed and wonder if the survivor needed this money. At the same time, I would also question why the federal government just required GE to prove twice they could not find the owner of U.S. bonds while the requirement for GE stock was three times instead of two.

Those days of paternalism were probably best illustrated by the way we promoted job loyalty among retirees as well as employees. Photographers Lou DiCerbo and Mike Decata and I were often on the job finding retirees at their summer camps enjoying the good life on pension. There were so many of them in the old days that you did not have to contact someone at a lake. You just drove up to a lake and asked a marina owner or storekeeper if he knew a retiree and invariably you would have three or four names with locations. We would case them out to assure ourselves they were indeed having a good life but even if the camp was not a high-end one, all you had to do was photograph them by the lake relaxing in an Adirondack chair.

There is probably not a lake within 200 miles of the GE plant that I haven't been to and I met some wonderful old timers who delighted me with stories of GE electrical pioneers. These unannounced visits sometimes did cause us to leave in a hurry at times and DiCerbo got pretty good at anticipating this when he noticed how young the wife looked.

Dealing with me had to be tough on our photographers, too. On one occasion I persuaded DiCerbo that since we had a truck clearly identifying us as GE, why didn't we knock on the door of the multi-million dollar mansion on Lake George owned by the Albany political kingpin everyone was raving about and see if we could get in to see it. He stayed in the van and at the door I was told they did not have a GE Refrigerator but he would let us check out the one from Whirlpool. Lou quickly drove off because I had identified him as the repairman.

Anyhow, the GE News life was about to change for me. In 1968, the general manager of the Large AC Motor and Generator Department, who I had interviewed in one of the state of the business reports, found out what I made as salary and he offered to nearly double it. I quickly left the GE News.

MOTOR YEARS

It was Bob Smith, an electrician who had worked his way up to become the General Manager of the newly-created Medium AC Motor and Generator Business, who called me and bluntly asked: "How much money do you get a year?" When I told him it was $7500, he said that even he was surprised it was that low. He immediately offered me $12,500 which took my breath away. The next thing he said was, "Never answer that question directly again. I am told other department newsletter editors get that amount and you can expect a raise in six months if you perform like you have been. Tell that Carl Hudson *(plantwide relations manager)* I just hired you. He should be ashamed of himself – particularly since employee relations people are supposed to be promoting civil rights legislation instead of violating it."

During the late 60s, "conscious raising" was a term most women were becoming accustomed to thanks to Gloria Steinem and Bella Abzug in the academic and political arena. Not only was Bob Smith a mentor to me, but soon Helen Quirini, a woman winder on the motor assembly line, also opened my eyes to inequality. Jim

Warren, Manager of Manufacturing, taught me the difference between appearance and reality.

The Civil Rights Act of 1964 had been thought of as an equal employment opportunity for blacks and GE had responded with an upward mobility program. But smart managers like Bob Smith also cited the equal opportunity and pay rules for people like himself who had the skills but were not considered high-potentials. Smith did not have a college degree but he had more than proven himself as a foreman, supervisor and manager of manufacturing. He was quick to tell me that in the future I should remind employee relations people they still would not hire Tom Edison because he was deaf or Charles Steinmetz because he was a humpback. He even added that most people working in employee relations had been put there because they couldn't cut it in engineering or manufacturing.

At the time, there was just a small and large ac motor and generator business. It was apparent GE did not have the market share lead in medium size motors and generators. So some 2500 employees were assigned by Vice President Oscar Dunn to Smith to make inroads into this very competitive sector. Smith was under pressure to create an employee relations section as all departments did under the decentralization philosophy of GE CEO Ralph Cordiner in which each business was measured by its own results and not as a part of another group.

To help his bottom line, Smith intended to delay setting up an employee relations section as long as possible. The business had over 30 competitors and he was determined to show he could more than compete with the best of them. While he felt he could out-manage anybody in the business, he knew he had a problem writing good letters and reports. So he told me he expected me to help out with that. I was elated, particularly since it meant that I could sit in on his staff meetings with the managers of finance, manufacturing, engineering and marketing. I was even more thrilled when he said he knew I could do it because he had read all the interviews I had been doing with department heads throughout the plant as Editor of the GE News.

He confided in me that he was going to delay as long as possible setting up an employee relations section to save costs and that work he needed done in hiring and dealing with the union would be farmed out to the SAC (Small Motor and Generator). He was being told by plantwide relations that a proper ER staff required some ten professionals to handle relations with white collar, clerical and factory employees. One of my jobs would be to convince plantwide relations people that I knew he was working on setting up a section.

Now fully realizing how I had been taken advantage of by plantwide employee relations, I had no problem in deceiving them at monthly meetings all communicators were called to with the plantwide relations manager. I described the new office we were creating for the Manager of Employee Relations and his staff. I told my friends in union relations it was just a matter of time when jobs would be created in that area, that Mr. Smith was pleased with the union relations support he got from SAC and that in addition to the advice he was getting from plantwide relations, he was also looking there for experts who knew the business.

The motor businesses in the plant were always considered second-class by its big, highly profitable Large Steam Turbine-Generator Department. If there was a problem with second and third step grievances filed by the unions heading for corporate mediation, Carl Hudson's plantwide staff would point to the motor businesses as causing labor unrest. Even the Industrial Clinic, which was part of plantwide relations, was found to be assessing businesses other than the reigning steam turbine department a dollar more an hour to take care of those needing medical services. It was suspected the assessment for plant security and the fire department were not fair but that was hard to prove because it was not based on employee visits. We used to joke that it was ironic the plant train ran through the plant at a point where it separated the motor and turbines businesses giving added emphasis to the motor business being on the wrong side of the tracks.

When I started working for Smith, he placed me near the manufacturing manager so I would learn more about the business. He suggested I set up offices for the employee relations staff he was supposed to hire in a huge spare conference room next to his office. I was to furnish it slowly and as cheaply as I could. I knew there was a company manual for how offices should be set up according to management level. But I figured I could save a lot of money by going to local furniture sales instead of ordering through the company catalog. As it turned out there was a furniture store going out of business and I got a terrific desk, chair, credenza, bookcase and couch for a little over $400.

Smith looked at the catalog and immediately gave me one of his Accent on Value *(at the time the company's new campaign as part of the "Progress Is Our Most Important Product" was called "Accent on Value")* dinner for two awards. I had saved $6500.

But when Carl Hudson, anxious to build his empire of relations managers in all the departments even further, came down to see the offices, he was upset. The furniture was better than anything he had. I was chastised for not knowing the office protocol procedures. But Smith quickly solved that problem and notified Hudson he had exchanged his furniture for that of this soon-to-arrive relations manager. So I got another Accent on Value award because he found the huge leather chair extremely comfortable.

Obviously, I was totally devoted to doing everything possible to make our motor business the most competitive in the U.S. In addition to a newsletter I called the Current News *(even though it only came out once a week)*, I also wrote letters, reports and speeches for Smith, put out news releases and, the most fun of all, handled the motivation programs. Smith was a huge believer in making motor and generator people feel like a family pulling together. At the time, the corporate recommendation was $5 per employee each year should be taken out of net income and spent on motivation programs.

In 1970, the Company-wide Accent on Value program, which had a ten year run under GE Board Chairmen Fred J. Borch and Gerald L. Phillippe, was augmented with a year-long campaign called "Best Buy." While the main street separating the power generation and the motor businesses was renamed Best Buy Boulevard, it was felt a plantwide program to recall the spirit of the early U.S. founders would ease the pain of reduced factory wages under the Make Schenectady Competitive program. So a "Spirit of the Seventies" motivational program was introduced in the Schenectady plant. Our Large Steam Turbine brother across the tracks announced an open house to show the families of employees how their jobs had electrified the nation. They also rented the local fairgrounds to have picnics for all three shifts.

I was challenged to come up with a program to also make our motor employees proud but to do it for $2 an employee, not $5. Smith quickly pointed out it was too costly to shut down production for an open house *(besides, our buildings could not compare to what was called the Grand Canyons of the Large Steam Turbine-Generator factories)*. A picnic also meant too much lost production time and there was not enough money for food.

So I put money into awards of beer mugs, pens and even Continental coin replicas celebrating the Spirit of the 70s. I also created displays of how customers were using our top quality motors and generators to make America great and put these up in areas common to office and factory employees. To give more visibility to the motor and generator businesses, I borrowed a street-full of life-size statues of Colonial soldiers from the mayor of Schuylerville near the Saratoga Battlefield. He was pleased to loan them to us because it meant publicity for them. These were lined up in formation on the motor side of Best Buy Boulevard.

Then I invited several area high school bands to march down the boulevard playing Yankee Doodle and other spirited rally songs. The students were more than happy to freely show off their talents.

Everything was going great and I figured I was in for another award if not a raise because I had come in at $1 per employee. But then during the Yankee Doodle band parades, Hudson called Smith to complain about the disruption. I had not realized most of those high school students had parents that worked at GE and many of them were in the Large Steam Turbine Buildings. They left work to see their kids marching by the motor buildings. Even though I tried to explain that they, too, had been energized by the Spirit of the Seventies, it was to no avail. Steam turbine had lost more production time than it had planned on.

Smith merely smiled when he hung up the phone. He reminded me that as long as you make money, you are safe at the top so I was not to worry. A week later he had even a bigger smile on his face when he told me that I was one of 50 winners throughout the company of the coveted Borch "Best Buy" corporate headquarters award. That was a form he filled out himself and sent in directly to headquarters without consulting plantwide relations. I still have the engraved clock with each hour set off by an artificial diamond developed by the Company's Research and Development Center.

Working for Bob Smith was probably my most enjoyable GE job. He was determined to show the company he could run the most profitable motor business in the company. He exposed me to every facet of the business by letting me sit in on his staff meetings. This also allowed me to do a better job writing speeches for him at the annual companywide General Managers meeting in Florida. I even filled out his corporate management development forms and tried to accidentally drop the section on academic background.

But he grabbed it and inserted a big "O" with an arrow pointing to the bottom line results. He often told me to ignore stupid advice from groups like plantwide relations which was trying to build an empire of its own. "As long as you make money, you are on the safe side with the folks at the top. So when you know you are doing a good job, forge ahead."

It was one of these times when I asked him if there was money in the budget to fix up an old car I had seen in the plant dump. I explained that plant security had been forced to lift the fake pass I had when I left the GE News. I even told him how I obtained that pass and since I was totally persona-non-grata with plantwide relations, I was not likely to get a real one. There were more than 15 motor buildings scattered on the south side of the plant and it was difficult getting to all of them by going to different outside parking lots nearest them.

However, my friends in Interplant Trucking had told me it would only take a couple of hours for them to get the old company car I had spotted in the plant dump back in operation. Smith called them on the phone for more details and true to his cost-saving nature, he asked me if I minded driving a car without a right fender and a bent bumper. I said "of course not," so the deal was done for about $50 and I had a company car. I have to admit I used to park the rusty old thing near Carl Hudson's parking spot when I could. It irritated him even more when I pasted a huge GE Monogram and Accent on Value sticker on it.

Smith's drive to save costs and improve market share by being more price competitive and introducing improved versions of the motor and generator line, were so successful he was asked to also head up the Large Generator and Motor Department. His mission was to get the competitive edge in the growing hydro-generator market as well as serve large industrial customers such as steel mills. The hydro-generator units now being built in the new MSC Bldg. 60 were actually larger than the 1,000,000 kilowatt steam turbines being built in Bldg. 273. I remember one of the largest units being shipped to the Glen Canyon dam in Colorado when some of the employees, during lunch hour, actually painted a scene of the canyon on the side of it. (*Years later, I was on a rafting trip through the Grand Canyon when I learned the guides, who were mostly environmentalists, did not share my enthusiasm about the hydro-generators, even though the water releases did make for greater white water trips.*)

As head of this business with major competitors competing for the large AC and DC motor business as well as hydrogenerators, Smith inherited an employee relations staff so he was no longer under pressure to form one. But my job did not change and I stayed in my office beside the manufacturing manager.

Smith always saw his challenge as motivating employees who were paid the same under the plant-wide day work plan to perform above and beyond the call of duty. Each staff meeting he asked his section managers for reports on how many suggestions had been turned in and how many had received monetary awards *(and pointing to me)* recognition even though he always read the newsletter. Again, I was reminded never to let promotion of the Suggestion Plan lapse because it was a way to reward those who worked harder and smarter.

We would try to save money on creative motivation programs. Instead of costly open houses and picnics, we sponsored splashy programs such as trips to visit customers which involved a different group of production and office employees each time.

The visits to customers became very popular. The one I remember the most was under our "LM&G Great People Care about Customers Program." Fifty employees were chosen to visit the Atlantic Richfield Point Breeze Refinery near Philadelphia where they were able to see motors ranging from 250 to 13,000hp in action, most of them GE. To promote customer trips, I had gotten a good deal on a 60-foot long heavy plastic banner from a local house painter to paint the message of "LM&G Great People Care about Customers!" in huge letters. We draped it across our largest building and it could be read from the highway. We rolled this up for customer trips and would unveil it for our customers with everyone standing behind it. The bus trips also included entertainment. In this case it was a trip to see arch football rivals University of Pennsylvania and Princeton. In our enthusiasm, after posing behind it with our customer, we also took the huge banner to the football game and unrolled it during halftime.

That was to get me into trouble with plantwide relations again. We were thrilled when the TV camera scanned us behind our huge "We Care about Customers" sign in the stadium. But unknown to me, the commentator suggested that GE must be getting desperate to look for free advertising at football games. But again, Smith said, "Ignore Hudson. We're making money and that's what counts."

Not only was my self-confidence being raised but so was my conscious thanks to Helen Quirini. I had met her earlier when I was on the GE News. I had done a story on a factory woman who told me what an honor it was to work in the factory and how much she loved her job and the benefits. Helen had come into the office, called me an idiot college kid and asked me if I had the guts to come down to the small motor department with her and look at the real world. I accepted her offer and it was then I found that women motor winders were getting 75 cents less an hour than their male counterparts. I asked about the union allowing this and she laughed. She pointed out to me that the company got away with that because New York State had a law on the books that women could not lift more than 50 pounds. So at the end of each line, a stator had to be lifted which weighed more than 50 pounds. Yet it had nothing to do with the job itself. In fact, she pointed out the women motor winders were even faster than the men because of the dexterity of their hands. I commiserated with her about my own lack of equal pay and absorbing two other jobs.

Now that I was in the motor group and closer to the action, I became friends with Helen and her black friend Edna, who thanks to Bob Smith, was on one of our customer trips. I found out Helen had been a leader of the United Electrical (UE) workers union in the 50s which, unlike the International Union of Electrical Workers, allowed blacks and women in their union. During the McCarthy years, the UE union was charged as being communists. *(Years later I would listen to tapes obtained by a University of Albany professor under the Freedom of Information Act which had Helen testifying before the committee that she was not a communist, that all she was doing was trying to get equal pay. When asked if she knew Karl Marx, she had replied, "I never met the gentleman. Is he here?)*

Edna and Helen got me involved in the Refreshing Springs Day Care center which was trying to help black and other poor women from Hamilton Hill by taking care of their kids while they were at work. In doing a newsletter for them, I soon found out what Helen meant when she showed me what a farce it was when it was said that women did not need as much money as men because they were bringing spare money home to their families. In fact, most of the women on the assembly line, black or white, were the sole breadwinner of their family. "Father GE does not know best," she used to say.

Whenever Helen got the chance to open my eyes, she gave me a challenge. It was not long after Governor Nelson Rockefeller built the huge government center in Albany which rivaled the one in New York City, when Helen invited me to go with her to an open meeting with state legislators. We were in one of the heavily-carpeted and marbled buildings to attend hearings about equal pay and opportunity for women. When Helen got up to plead for bargaining rights for household workers -- which included the right to participate in Social Security -- I stood up with her and a group of other women friends she had organized for the effort.

Standing up to plea for equal pay for GE women and for legislators to get rid of the lifting the 50 pound rule was the leader of a group of maids who worked and often lived in the mansions of their Long Island employers.

The reason for switching roles was the fear of retribution. If a picture were taken, it would be hard to find the culprit. The maid from Long Island did a great job challenging the legislators to lift a series of barbells with her to show women winders could lift more than 50 pounds. Helen was great as the Long Island maid who took on one of the Long Island women who testified against paying Social Security. The elegantly-clad lady ended her talk by saying: "We love our maid...We even take her on our family vacations, sometimes to Europe." Helen noted that it was not a vacation because maids would have to sometimes take care of at least ten kids in the family clan while her own kids had to fend for themselves.

In asking for Social Security participation, Helen declared these employers could certainly afford to match the contributions and wouldn't it be nice for them to know they had helped provide a retirement for their workers -- especially, she said, because of the way they are hired and fired. She told of suffering the indignity of being picked out of a long line of household workers on a stage in which the strongest and youngest were selected for household work. The older and weaker ones were sent back to the line to be paid a cheaper rate, if hired at all.

I will always remember how the well-dressed legislators and Long Island ladies, situated in those posh surroundings, did not intimidate Helen. But mostly I remember her chewing out a protester who had stubbed a cigarette out on the lush carpet. She railed at her for destroying property that all of us taxpayers owned.

Thanks to the bra burners as well as the Martin Luther King marches, employers like GE were paying attention to the changing times. To meet the 1972 Equal Employment Opportunity guidelines, orders came down from the Vice President of Human Resources at headquarters to also start hiring women foreman (*there had been many during World War II but when the men returned these jobs were no longer open for women.*) Naturally Carl Hudson ordered the motor department to place the first woman foreman who was being sent in by headquarters. So as Manager of Manufacturing that became Jim Warren's job.

Warren, like Smith, already had his problems with the plant-wide relations when it was found a black man with a college degree was working in his stockroom. His promotion to foreman came right after the Civil Rights Act of 1964 and Warren was still apologizing for not knowing that Eddie had a college degree.

Eddie was now one of his best foremen. It had been amazing how quickly he had gained acceptance of the 40 men on his shift. Warren had warned Eddie it was going to be tough but Eddie had correctly predicted that he had an advantage. He had been right. Thanks to Eddie's fast sure shots, the stator winding assembly group was now the best of over 50 basketball teams at the GE Athletic Association.

Warren often lunched with Eddie, enjoying his sense of humor and the teasing he got from him warning him that he just might have to mention to the VP of Human Resources that Warren was born a Mormon in Salt Lake -- and that they still did not accept blacks on their the Salt Lake team. "Obviously the VP does not read the applications of employees unless he is forced to," he teased.

In fact, Eddie had come to Warren's rescue when he and other manufacturing managers were told by a new naïve corporate employee relations manager from headquarters to improve union relations for the upcoming national contract negotiations by dressing up as Santa Claus and passing out cookies. The suits and cookies were provided with instructions on it to remind employees it was a season of receiving as well as giving.

Warren prided himself on being as tough as his workforce. He was reluctant to dress up as Santa, particularly since the huge suit fitted his rotund form so well. While as a Mormon he did not drink, he decided to brace himself with a six-pack he had taken from the night shift workers. He decided to start with Eddie's workers who had already shown their tolerance. It was here that he literally lost the cookies he was to pass out. Eddie quickly explained the situation and the workers agreed with him that the idea was not only ridiculous but insulting to their intelligence.

However, they did send a note to Hudson through their union that they really appreciated the treat and they were looking forward to even greater treats from the Easter Bunny. The serious suggestion of how about starting the New Year with a pay raise was a PS note.

So it was natural Warren asked Eddie for help when the new lady general forewoman was to arrive. Eddie read the formal letter on "Upward Mobility" stationary the headquarters VP had written to introduce Amelia Stone. She was a graduate of Vassar, specializing in women's studies. A copy of her Phd thesis, "Conscious Raising of Men in the Industrial World," was enclosed. She was highly recommended by Corporate for her ability to gain acceptance of women in management. At the local level, Hudson sanctimoniously added that with her background she should be immediately added to the high-potential list for promotion.

Eddie was there when she arrived in Warren's office. She was dressed in a three-piece navy blue pants suit with a huge white bow bustling out from the vest. Her hair was done in a tight bun and, without being asked, she immediately sat down in a rigid position. Eddie complimented her on her academic background and then commented that she must know the environment she would be working in because even the men with longer hair pinned it back to keep it out of the equipment. When he suggested that she might want to tuck in the huge bow to keep it out of the automatic winding machines, she bristled, "This bow is my signature as a woman. I don't intend to get near the machines. I will be managing the men who should be at the machines, not me."

When Eddie volunteered to introduce Amelia to the stator bar section employees, Warren was grateful. He knew Eddie would require a free lunch for it but it was a great relief not to have to figure out how to be politically correct. As Eddie and Amelia were approaching her factory office, which Warren had cleaned up, including the installation of a new desk and chair, Eddie noticed that Warren had forgotten one thing...that old pinup of Betty Grable was still on the bulletin board beside the OSHA regulations. Amelia immediately went to it and ripped it off despite Eddie's protest that perhaps it would be best if she asked the men to remove it and explain why.

But Amelia only glared at him and said, "There are certainly situations where you don't ask for something you might regret. What if they said they won't do it?"

Eddie had to admit she was probably right about that. He suggested she not leave the poster in the trashcan beside the bulletin board. But again she challenged him replying that the men needed to know such pinups would not be tolerated. As the men coming in for her shift punched in their time cards beside the bulletin board, Eddie could see they were in no mood to deal with a woman, much less one who was messing around with their idea of what a woman should be like.

Eddie gave them one of his "time out" signs they used in the league, hoping they would chill out over the posters destruction. He then introduced Amelia to them. He was relieved to see there were just stony stares. No one stepped forward to retrieve the poster. When the buzzer rang, they went to their work stations and Amelia disappeared into her office. It was only twenty minutes later that Amelia rushed into Warren's office, shouting, "Of all the nerve. Of all the nerve!"

She explained she made her rounds of the assembly line to assure herself production quotas were being met when one of the guys exposed himself to her. "I can't believe the nerve of that guy. I want him fired now."

Warren explained she had to file a grievance according to contract negotiations and gave her the phone to confirm that rule with Hudson. Warren did not know what Hudson said to her but he was not dismayed to see her leave, slamming his door in the process.

Again, Eddie came to the rescue as acting foreman for her area as well as overseeing his. But most important to the frustrated Warren was Eddie's suggestion they hire a woman he knew from the motor winding section in another building. She was also the archery champion at the GEAA.

After a few phone calls, Warren, knowing that an EEOC official was coming to the plant the following week, now had an agreement to quickly hire a candidate he had for the job. Her name was Sandy and when she arrived in Warren's office she was wearing blue jeans, sporting safety glasses and made herself at home by helping herself to one of his cigarettes. When Warren said he would take her to her new office, she said she had already been there and she did not need an introduction. She knew many of the assemblers from the GEAA.

Warren relaxed and waved her out the door with a smile. He figured Eddie was right. She might not have a degree in cultural relations but he figured she was up to the job.

His confidence was confirmed at the end of the week. She had even gotten the men to exceed their production quota. He found out from Eddie that not only had Sandy put back the Betty Grable poster but she had added some graffiti in the form of a rifle pointed out much like the traditional Uncle Sam recruiting poster. She had written under it "Quit staring at me and get back to work."

As for the guy that had exposed himself to Amelia, he had been quickly put in his place and laughed at by the men when he pulled the same stunt on Sandy. She had grabbed his penis and zippered him back up, noting that with a small pistol like that it was a wonder it would fire at all.

Warren was pleased to report to Corporate Employee Relations that Sandy was a perfect general forewoman, having improved productivity in that area by 5 percent the first week. Still, at headquarters they gave Amelia credit for breaking the glass ceiling for other women aspiring to be a forewoman. As a high-potential she would be promoted to manager of union relations.

While the promotion disgusted Warren it didn't bother Eddie. He just laughed and said the experience may have taught her a thing or two so when the shop stewards met her at the negotiating, table they might just remind her of the incident.

Warren decided he now had a replacement for himself when he retired. He put Eddie's name on the high potential list.

What always amazed me about Warren was also true of Ginn. Both men had more money than they would ever make at GE. Ginn's wife had been the daughter who had inherited the Colt fortune. Warren was a Mormon who inherited several thousand acres of mountain land near Salt Lake City back in the 50s. Today, there are several ski areas there including the Park City Ski Area. He used to show me the checks he received each week for chalet parcels.

I decided they loved producing products and doing business. Greed was not their motivator but as Smith would say, A Fair Day's Work for A Fair Day's Pay is a good motivator for management as well as the workers. He was into rewarding those who performed above and beyond the call of duty and he rued the day piecework had ended. He found ways to reward workers through the suggestion plan and the customer trips.

Smith also didn't suffer fools. I remember his disgust with an OSHA inspector who simply had to find something wrong at each inspection to prove he was doing his job. To save costs, Smith used to have a few inexpensive problems in place for him to cite us with such as a full trash can with flammable material placed near the welding station. But one time, he couldn't resist giving out this order when the OSHA inspector informed him that all ladders in the motor plants which were over six feet had to have protective steel cages around them to protect the workers going up them. Smith called in Warren and ordered him to chop all ladders in the business down to six feet.

When a vice president was sent in by the Industrial Group to assume a position of authority over the three motor businesses to compete with the steam turbine vice president who had plantwide authority, Smith was not impressed. The guy had no experience and was a bit of a stuffed shirt. So Smith told him that before he retired.

By this time, the Equal Employment Opportunity Act was opening new doors for me at headquarters. But first, a review of why we were glad there were some women out there burning their bras.

NO THANKS

In the sixties and early 70s (*before EEO and my getting older*), the best way to avoid unwanted advances often took some interesting evasive actions but one of them proved to be a lot of fun.

It was during contract negotiations with GE unions that I and four editors of department newsletters were asked to come to the Biltmore hotel in New York City to publish a companywide newsletter. It would be sent out to some 150 plants across the U.S. by fax (*We didn't have internet then*) and the quicker the better -- because the union side also had their communication team to reply to the company offer.

It was felt if the company was out there first with its offer and presented it in a favorable way, the union members might decide to vote for it or at least pressure their union leaders to accept the national contract offer.

The four editors going to the city with me were people I knew -- good, supportive friends from our days of promoting the Make Schenectady Competitive campaign at Jimmy's. When all of us arrived that Sunday for the coming week's negotiations, the first thing we decided was we would go to the bar to watch a football game after we got settled in our rooms. When I got to the bar, I looked around for them and decided the guys must still be getting situated since they were rooming together, So I was about to sit at the bar when a large, burly but well-dressed man *(even bouncers were well dressed there)* came up and told me to leave. Despite my protests that I was not soliciting there but had GE friends I was meeting, he roughly ushered me out a side door.

I found my way out of the alley and went back into the hotel where I again approached the bar figuring my friends must be there by this time. When the bouncer saw me, he interrupted me before I could get to their table where they were watching TV. Again, he had me by my arm as I yelled to my friends. Quickly sizing up the situation, they decided it would be a good joke to say they did not know me. However, when they saw I was going to be roughly ejected, they came to my rescue.

All of us had a laugh over the incident. They decided that I should skip a round in paying for our drinks since I had been so rudely ejected. In fact, they tried to get the bartender to give us all a free round but he said I should have shown some GE identification when I came in or it would not have happened.

Still, we had a good time that evening. Between toasts and touchdowns we wondered how we were ever going to get a newsletter published as fast as management wanted it out. Ted Evenden had been on the Corporate Employee Relations communication team before during negotiations. He complained that lowly communicators were the last to know what was being discussed as senior vice presidents first confided in regular vice presidents who then met with their staff to tell them and by the time it got down to us, a day was lost. They would forget about the delay in the chain of command and then complain about the slowness of getting the word out to the plants about the terrific offer the company was making to the unions. Ted, draining his glass, declared that even the secretaries, typing up the negotiations from the prior day, knew more than communicators.

It didn't take long for all of us to realize that if I could be mistaken for a solicitor, why not a secretary. So the next morning, I found my way to the outer room of the negotiations center where I did present my credentials. There were three secretaries from corporate so I introduced myself as a proofreader, handing them the Schenectady GE news with my name now on it.

They went out of their way to welcome me. It seemed that in a prior year one of them made a mistake in typing and actually gave the union an extra percentage in medical benefits. The cleaned typed copy was quickly agreed to by the union when they realized it had not been noticed by the corporation union relations staff. That typist had been fired and they were warned that it should never happen again.

They were not only pleased to have me join them, but also relieved when I told them I would take full responsibility as the final proofreader of everything they were typing. So for the next four days, I enjoyed their company and the excellent meals they were served. In fact, because of high security, their suite was just off the negotiations room and they were not let out of sight. Their meals were catered. I often stayed after negotiations work for their excellent dinner meals as well as making sure everyone had left the negotiation room. Then I went back to my room where I met my fellow newsletter editors where we worked on our companywide newsletter. Afterwards, we adjourned to the bar for a celebration in which I was not to pay for any drinks.

So when negotiations ended, we had our newsletter ready to be faxed across the country just an hour after we were told. The company negotiators were amazed at the excellent quality of the newsletter and how fast we had produced it. We were invited to their celebration of a successful four-year companywide negotiation.

Here the bar was open and a smorgasbord of gourmet food was available. It was while I was trying to figure out what that black, beady stuff was that everyone seemed to be delving into when a Corporate Relations Vice President walked up to me and said: " I hear from the other communicators that you are mostly responsible for the excellent communications which went out to our plants. They couldn't say enough good about you. Congratulations. We are glad to have you on our team. In fact, I would like to give you a special reward later this evening. Here's my room number."

I was so stunned I just took his card. I had just taken a bite of the caviar on a cracker like I had seen the others do. He didn't seem to notice I dropped the rest on the linen table cloth as he sauntered off with a condescending smile. I couldn't believe my ears much less my taste buds. I quickly threw a napkin over the mess and looked for my fellow communicators. I soon found them at the open bar and immediately told them their good intentions of praising me was now a major problem. They got me into this mess. Now they had to figure out how to get me out.

They were as offended as I was with this turn of events. We discussed hiring someone to take my place but decided that even a professional should not have to put up with such a hypocrite. Besides, none of us wanted to come up with the money to feed his ego. Then it hit us. We had planned on enjoying the night life of New York before we left for Schenectady the next day. So why not stay out all night and come in to the last breakfast meeting raving about the all-night places in New York City.

My four friends even made sure the Vice President heard them call to me that our dinner reservations in Greenwich Village were confirmed.

It was a night I will never forget. We did start at Greenwich Village but it was at a bar, not a restaurant. I was so grateful for their help, I offered to buy the first two rounds. I also knew these guys all had wives and kids. In fact, Ted was the one who had been a negotiations communicator twice before. He told me he saved a kitty just for this once-every-four-years negotiations occasion. He had six kids and I knew one of them would be starting college soon. So when he flashed $200 in front of me, I was determined to see that he did not spend it all.

That turned out to be a tough job. He was always the first to throw out a $10 bill. But it got easier as the evening wore on. I was able to scoop up his $10 bills without him noticing. We were in a place called the Bookshelf where the dancers gyrated on shelves that went across the wall behind the bar. The bartender, noticing Ted was entranced with them, told us they put the bookcases behind the bar to make them less accessible to drunks. In an attempt to attract their attention, Ted was dipping his comb in my ale to slick back his hair in the style of the Fonz. It was suggested by the others -- and by this time I readily agreed -- that I should be drinking Scotch like they were since it would kill his dandruff.

When the place closed at two a.m. we went out to look for another bar. Again, the group came to my rescue when we saw a rough looking gang coming down the block we were on. You could tell they immediately sized us up as tourists as they commented on the suits and ties the guys had on. I tried to hide among them but they spotted me in my dressy suit. I don't know whose idea it was, but it was one everyone quickly agreed to when it was decided the gang was more interested in me. Fred and Dick would go back, cross the street and around the block to see if they could find a cab or get help while Ted, John and I would cross the street to see if we could avoid them. It turned out to be a wise decision because even though we crossed the street to avoid them they were right behind us. We even began running to meet up with Fred and Dick who came toward us yelling that a cab was on its way. As luck would have it -- and were we ever grateful for it -- a police siren sounded several blocks away. Then it was the gang running, not us.

While we were congratulating ourselves on being so smart, Ted noticed we were standing in front of a place called the "Body Shop." Ted began laughing when he opened the door which led up a set of stairs lined with photos of the bodies of naked women. I was reluctant to go in but they assured me they had proven to be great chaperones. So up the stairs I went with them. At the next door, a bouncer asked for $20 each. When he saw me, he told them there was no charge for women and laughed. He assured me that I would not be part of their show.

So in I went. To my amazement, we entered a plush room containing a very fancy circular stage. The bouncer seated my friends in the first row and seeing that I was concerned when he motioned me not to follow them, he told me I could sit right behind them which I did. The first act was a comic who came out flipping a deck of cards which he began to sell after telling some raw jokes. Ted was the first to buy a set when he advertised 52 cards with 52 different positions. Next came a bevy of beauties who advertised lap dancing. I was totally naïve until I saw what was happening to the customers in the front row. I couldn't believe it. It was then that I retrieved a bunch of Ted's $10 bills when I saw the lap dancer reaching into his pockets. I was sitting behind him and got to his back pockets first.

My friends decided it was time to leave when it was announced the highlight of the late night performance was Sally the Shepherd, a huge ewe which was brought out on stage. Again, I was stunned when I saw what was going to happen. It was too much for our group, We quickly left, found a bar a safe distance away, and counted ourselves lucky the place was not raided while we were there.

When the sun began to come up, we went to the park to plan our strategy. They would have me by the arms when we came into the meeting room where breakfast had been set up. They would note how I couldn't hold my drinks and had passed out. So they just let me sleep in an all-night movie we had gone to. There was some shaking of heads and the vice president in question looked disgusted. On the way back to Schenectady by train, we had quite a laugh and another round of drinks. Ted was even happier when I produced $120 of his money which I had confiscated, most of it from the delving hands of a lap dancer.

But avoiding advances for some women was not as easy. One of our plantwide relations managers (*Bill, you know who you are*) was constantly hitting on his secretary/stenographer, Kathy. She was very good-looking and certainly not interested, but she needed the job and was supporting her ailing mother.

When he called her into his office, he would make all sorts of suggestive comments while she was taking notes. When he got up to supposedly admire her shorthand, he would put his hands on her at which time she would find a way to hit the wall to signal me for help. My office was on the other side and she always told me when she was going in to take dictation. This was my cue to interrupt the session. I would rush in often announcing that Kathy had an emergency call from her mother, hoping it would also perhaps tweak his conscience.

However, when we were having a Christmas party at the Van Dyck in Schenectady in which spouses were invited, I inadvertently did Kathy a favor. The rumors had gone around that Bill had married for money and that his wife was also several years older. None of us had met her.

I was at the bar getting a drink when this well-dressed, prim woman sitting there asked me about Kathy. She suggested that she looked like someone who was into extra office activities. I got the gist of her comment and quickly told her that on the contrary Kathy's good looks made her a target, particularly for our manager who was as sleazy as they come…that Kathy was as smart as she was beautiful…and that the last person in the world she would get involved with would be Bill. I told her about my efforts to save Kathy from his grasp to discover that I was talking to his wife.

After that I did not have to interrupt Bill when Kathy was in the office but I sure got some hard looking stares from him.

But undoubtedly the worst case of victimizing women came from a marketing communication manager. Like so many other businesses which got their start in product development laboratories in Schenectady, the Gas Turbine business was on a strict budget until it started making money like the highly-profitable big steam turbine power generation business. While motor business managers felt the employees they trained would soon jump ship for better paying jobs in steam turbine, it was the gas turbine business that struggled for market acceptance against its power generation brother.

During those lean years, its marketing communication manager (*I refuse to give his name because his wife is so nice*) often bemoaned the fact he only had a tenth of the money that his counterpart in steam turbine had to lure in customers. He used to say that all his steam turbine counterpart had to do was entertain a handful of utility customers with lavish dinners, hunting trips and a safari or two and for some customers they could easily afford showgirls for private entertainment. Gas turbine markets were more diverse and international with hundreds of potential customers, particularly in the Middle East oil fields.

I was editing the GE News when I first met him. As noted, I regret all those cheesecake photos I was putting on the front of the GE News to attract what my manager hypocritically called red-blooded, blue collar workers -- even though there had been times when he had vetoed some photos saying she was not that pretty.

I have to say that this front page feature became so popular that some women were volunteering for various poses, often coming up with their own reason for it. What I didn't know was that some of these women were getting calls from this marketing manager who told them he was a professional photographer and would like to take some really good photos of them. What he was doing was checking them out to see if they would be interested in helping him with his promotion of gas turbines.

I found this out when I was covering the arrival of some 15 Japanese customers at the Star Dust Inn who were touring the gas turbine plant. He had put together a state-of-the-art technology program for them in one of the meeting rooms which had been well received. The GE News photographer and I celebrated our success in getting many fine photos of the customers with the displays. I had some good quotes from them on the tour as well as comments on their favorable impression of the reliability and efficiency of GE gas turbines. We were at the bar celebrating this when one of the girls that I ran on the front page in her bathing suit came running up to me and asked me to take her home.

I left with her to find out this marketing communication manager had been paying her for her work with customers. She had three children, was an entry level secretary, and the pay for one night was more than she made in a week. She had even put on the kimono the marketing manager had provided for her but when her "customer" came in she couldn't do it. Her father had been killed in World War II.

The bar manager realized what was going on and reported it to the owner. Unfortunately another girl I had also run a cheesecake photo of was still serving customers when the police arrived and they arrested her. So in the middle of the night after I got her home, I got a call about her friend being in jail and could I help her.

Feeling somewhat responsible (*and today looking back, feeling disgusted*), I decided I needed to tell A.C. Stevens, the works manager about it. After all, when he found me on his couch that morning he had been very understanding and helpful. I went into his office when I saw his secretary was not at her desk and with some trepidation, I went in and told him the full story. He was shocked and immediately went into action, giving me money to bail out the girl and getting me to promise not a word of this would get out.

While the manager was not fired, I was assured he would not be soliciting low paid secretaries for marketing help. In fact, during this era of Upward Mobility for minority employees, particularly black employees *(the 1964 Civil Rights Act)*, he told me to tell my two white friends, both of whom he found to have excellent job ratings, that they were also eligible for the fast-track upward mobility program. Even though this would nearly double their pay, I did have trouble convincing the one with the World War II father killed by the Japanese to participate in the program. It turned out she was also prejudiced against blacks. But with her friend enrolled, she eventually agreed to sign up for the Upward Mobility program. *(Today, I am happy to say her mathematical and organizational abilities led to a high-level managerial position in assuring product components were shipped on schedule all over the world.)*

Fortunately, for this marketing communication manager, business did pick up and his budget was increased to entertain customers legally. Years later, I asked him how the professional showgirl business was going and he gave me the finger. But there were times when sleazy guys like him came in handy for the company. I remember when an executive from the Pakistan Water and Power Authority was visiting the plant to check on a huge $82 million order for gas turbines. He was housed in an elegant hotel room in Albany, NY. He had been there just two days when a call came in from the police that he had been picked up for soliciting a young boy in the park. They had him in jail. He was very indignant about it, saying that it was not an unlawful practice in his country and that if GE wanted his business they had better take care of him and quick. So this manager, who only hit on girls and was disgusted at the Pakistan customer, did know how to handle the situation. As he noted, there must be something to his our-culture-is-different story to the police because he had treated the boy to candy in the park at noon and then gotten his address to take him out that evening. When he got there in the GE limo that had been provided for him, the parents had the police waiting. Obviously, the situation needed to be handled delicately and could not be repeated so this marketing manager equipped the Pakistan customer with a collection of lurid films which kept him busy the rest of his stay. As he told me later, you never know what it takes to keep a customer satisfied.

In fact, one has to realize he proved valuable in meeting some customer desires. He was quick to suggest that state-of-the art seminars should be held in Amsterdam, Holland, where prostitution is legal -- and of great interest to the many Saudi customers using gas turbines to pump their black gold to the ports. In fact, when one of our more amorous guys going off to one of those international gas turbine state-of-the-art technology sales meetings gleefully told several of us he was going to have a great time making his selections for the evening, we gave him a pack of prophylactics and told him to use these so he would not give a disease to those selections.

Back in 60s and 70s, secretaries were often fair game because there was little protection from harassment. I knew several secretaries who knew more about the business than their bosses did. Even other managers would recognize this by waiting for their boss to leave and then ask them for information they needed.

On the other hand, I remember one great looking secretary to a general manager who used her position to lure young ambitious guys into her bed if they wanted preferred access to her boss. And then there was the manager with a job opening for a secretary who declined to hire a Certified Professional Secretary (CPS) from the Mildred Elly College but had the gall to note he was looking for a secretary with two outstanding points. I found out about it and sent him a note that his next new boss would be an aggressive homosexual.

I remember a guy in Bldg. 23 who used to wear a pendant chain around his neck. In the pendant he claimed to have the remains of a vasectomy he had to attract what he called liberated women, particularly those with no bras. He promoted his availability by declaring birth control pills had not been proven to be healthy for women. He also advertised he was a Planned Parenthood supporter.

Yes, times have certainly changed. Even today one cannot say that GE and other company employees act as a Normal Rockwell family despite discrimination and sexual harassment laws. In fact, I know one former lecherous VP of Human Resources who learned the hard way that things were different. He was set up by a female relations lawyer who taped his overtures on one of the company planes. She earned herself a million dollars to keep it quiet.

FITTING IN AT HEADQUARTERS

In 1972, when GE was one of four companies charged by the Equal Opportunity Employment Commission for discriminating against women, my job and pay prospects took a leap forward.

In fact, I was one of the first women from the huge Schenectady GE plant to be hired by Corporate Employee Relations at Fairfield in Connecticut along with a black woman from Appliance Park in Louisville, Kentucky.

We were given front offices in what some of those driving by on the four-lane Merritt Parkway used to call those two huge wedding cakes up on the hill. Our desks were even positioned near the windows so we could be seen.

Jackie's job was to monitor the company's progress in promoting females, white as well as black, and part of mine in corporate employee communications was to write press releases on such progress. I remember Jackie telling the EEO office they should measure quarterly reports by GE and other companies in numbers, not percentages. For example, if just one woman was hired in a professional office, the report could show 100% improvement. She survived that protest as I did when I objected to my full name being put on the press releases. It was just 10 years earlier that Lawyer Emil Peters had declared it was illegal not to publish an identifying nameplate on the GE News which resulted in me finally getting the recognition as editor. So this time, I was the one to insist my initials be used and not the full name. Both Jackie and I knew we were tokens but we decided there was a limit to being used.

We were visual proof affirmative action was working -- a label I hated, especially since my performance reports showed that I had never missed a day's work in 15 years and that in job after job I got excellent performance ratings. On several occasions, I had absorbed the work of two or three other people. My black co-worker, Jackie, felt the same about the label.

Still, we knew that our salary --and perhaps opportunity -- was finally coming up to what those, we called WASPS, White Anglo Saxon Protestants, were getting.

But fitting in was most difficult because there were very few middle income people among all the executives and their staff. When headquarters was moved from New York City, most secretaries did not make the move because they could not afford to live in Fairfield, Ct. Some were able to commute. The secretaries to the top executives were given administrative titles to justify paying them more than the normal secretarial rates. Pant suits had become fashionable among them, particularly the commuters. Jackie and I were advised to wear business suits to distinguish ourselves from the clerical workforce. Jackie was better at getting her bow tie tied and tucked into her vest. I finally gave up and wore a string of pearls which was acceptable.

Perhaps it was this attention to dress that got me into the first "let's have a discussion" situation. Because I was all over the two headquarters buildings interviewing staff for a monthly magazine called "Report for Managers" sent to plant managers throughout the U.S., I was named as head of a United Way drive for the Fairfield area. This became a dilemma for me. Coming from West Virginia, I knew there were no poor people in Fairfield County. When I found out the Boys and Girls Clubs already had several swimming pools, I successfully argued for the money to be sent to nearby Bridgeport which did have a needy population. While I was doing this, I decided I would also ask for used clothing donations because I was going to West Virginia the next month. Little did I know how swamped I would be with some of the most elegant, designer clothing I had ever seen. It took me three trips with my jeep to get it into my apartment. It was tempting to wear some of it to work. Still, I was one of the best dressed women in West Virginia for several years.

However, word of the clothing drive appalled the Vice President of Employee Relations Office as being undignified. Since the "harm" was done, I was able to keep it for West Virginians. It was then his attention focused on my driving a jeep which was just like the ones used by the maintenance personnel. By that time, a woman from Xerox had been hired who saw every other woman as competition. She preferred to be the only woman on the staff. She wore her bustling bow tie even better than Jackie and was quick to point out her job in corporate marketing was not EEO related. She told them I had painted the letters "Renegade" on my jeep and suggested it was difficult for me to shed my redneck background. Fortunately, the VP had a son who owned a Cherokee Chief. He knew Renegade was also the model of the Jeep and therefore I was not into Jeep graffiti. Still, he felt it was an inappropriate car for a headquarters employee to be driving.

I was not about to change vehicles since I still owed car payments plus the Jeep was handy for hauling my kayak for weekend river trips. In fact, I was still miffed at myself for forgetting I had it on top of the car when I came in to work after a tiring weekend competition. I had no problem getting it into the garage in our West Side headquarters building which was for staff including Grounds and Maintenance. As a result, the bay doors were high enough to accommodate trucks as well as Jeeps used by Plant Maintenance.

But, because I played bridge at noon time with the chef for the Executive building next door, I sometimes drove it over to the East Side Building if I was in a hurry rather than walk over through the long underground walkway. Forgetting about the kayak on top, I drove down into the parking garage entrance to hear a crunching noise because the kayak was not low enough to go under a door built to accommodate limousines. The chef, known to have direct contact with the CEO every day, interceded and kept my employee relations bosses from finding out about it.

In fact, if I did have a friend there, it was the chef. He disliked pompous people as much as I did. He often told me Board Chairman Reg Jones felt the same way but because of his position most people were afraid of being candid with him. Therefore, a lot of things were supposedly done in his best interest which he was not aware of. Probably the two biggest yes men who went a step further in exerting their power in his name were the Vice President of Corporate Relations and the Director of Buildings and Grounds.

When one of the maintenance men told me it looked like his boss, Elmer Toth, was going to be fired because the huge U-shaped bank of gold flowers outside the elevator leading to the board chairman's office were showing signs of turning brown, I couldn't believe it. He told me a fresh batch of yellow asters arrived every Saturday to give them time to get accustomed to their new environment for prime time Monday morning viewing. The air conditioning had gone off in the building over the hot weekend causing a few of the flower petals to start turning brown. I had just been over there earlier that morning delivering a Report for Managers news release draft and I hadn't noticed anything. I told the chef about it and he just shook his head, saying he would take care of it.

The Buildings and Grounds director was soon informed that Reg Jones had assumed the flowers were plastic since they were always in such perfect condition. In the interest of saving costs, he suggested it would be a good idea to have some nice artificial ones to save costs which would allow maintenance to concentrate on grounds work. He also wanted the maintenance crew to be commended for their superior work.

Our West Side building surrounded an atrium filled with perennials, a couple of trees and lots of shrubs including year-round greenery. Because it had not been that long ago that headquarters had been relocated from Lexington Avenue in New York City, landscaping for the new buildings had been a challenge. The grounds men had been told to plant life-size trees and mature shrubbery around the buildings to appear as if it was natural and not in a nursery stage. The Grounds and Buildings Director was not about to tolerate any aberrance. When the leaves on some of the new trees did not fall in October as they did on trees surrounding the area, he ordered the leaves stripped from the trees. This, too, was done on the weekend to avoid any questions. But because I was often there on weekends to make sure everything I was doing was the best job possible, I asked what they were doing. The chef and I had a good laugh about it, figuring the crew probably liked the overtime.

While I realized being at headquarters required conducting oneself with some decorum, there were times I simply could not conform. When I was walking the beach near the Milford, Connecticut apartments where I lived, I noticed a very dead and large horseshoe crab. I decided to send it to Lee Tomlinson who worked at the Research Lab in Schenectady to get even with him for sending me those huge mystery blueberries in a medical pillbox back in the days of the Make Schenectady Competitive campaign. Plant security had confiscated them and sent them away to be tested for poison while they conducted searches in all the mailrooms to find the source. When Lee, at a ski club meeting one night, mentioned the berries, I immediately called off security who wanted his name because of all the time they spent on the case. I refused to give it.

Sending the horseshoe crab via corporate mail to Lee turned out to be a dilemma for him, too. Instead of regular mail, corporate sent it by truck since it was so big figuring it was something important for the Research Lab. Then Lee had to refuse to say who sent it.

But there was one prank which I certainly do not regret today. When the Gerald L. Phillippe awards were given to employees who made outstanding contributions to their community, I remembered my friend Helen Quirini. She was certainly one of these with all her work helping to create the Refreshing Day Care Center for children of working mothers as well as an advocate for civil rights. Her nomination had been sent to the Schenectady Plantwide Relations manager Hudson *(the one that called strikers animals)* by the Large Motor and Generator department. But her entry never arrived at Corporate. So I got a copy of it and dropped it off in the in-box there. Helen was named one of the 10 winners in the company that year. No one in Schenectady ever knew how it happened and the relations manager evidently decided not to say anything about it since it was already announced by his corporate boss.

Today, I am also pleased that my respect for engineers caused me to want to help them with their negotiations with Corporate Union Relations. During the late 70s, the engineers and technicians at the Knolls Atomic Power Laboratory (KAPL) operated by GE in Schenectady were trying to organize for better working conditions and salaries. The threat of a white collar union had always been of great concern to GE. In fact, a movement to get rid of all the GE athletic associations across the company was underway to avoid contact between factory and white collar workers.

A skiing friend Ernie Johnson was one of the KAPL organizers. He knew most of the engineers were too conservative to form a union. But he also knew the threat of a union might move the company into meeting some of their needs. So between the two of us we were able to pass along enough credible information of their organizing activities to Baldwin, who headed Corporate Union Relations. This resulted in him advising KAPL management they should institute some of the suggestions the organizers were making. The vote for a union never came but some overtime pay and improved benefits did.

In fact, my only regret looking back was delaying an operation until the Christmas holidays so I could meet a deadline for a Report for Managers. I needed to have a tumor removed. It turned out the tumor was benign but medical treatment during holidays is dismal. That is a time when many doctors and nurses want to be with their families. Complicating the situation was a major snowstorm in Boston and Connecticut that dumped 30 inches of snow and was dubbed "The Blizzard of 78."

Connecticut Governor Grasso declared a state of emergency. I was coming out of surgery in the New England Deaconess Hospital when the first foot of it was on the ground. I had been placed in the cancer ward and could not believe the suffering some patients were going through when the people who were authorized to give morphine could not get to work. My roommate, a devout Catholic with incurable lung cancer, begged me to unhook her from the oxygen tank. I was actually considering it when a priest came in to give her the last rites and hear her confessions. I was amazed at what a decent and sacrificial life she had led. So when she said she wanted to end it all, he told her that she was sinning. She began crying. He gave her a couple of Hail Mary's and was about to leave when I got into a ruckus with him about calling me names.

Unbeknownst to me, I was bleeding badly internally because I had running around dragging my tree of bags giving me antibiotics and pain killers behind me. I thought he was saying Hell Mary, not Hail Mary. Then I got into an argument with the priest about telling such a wonderful woman who had lived such an exemplary life that she was sinning.

An intern heard the ruckus and it was he who saved me. My surgeon had left on vacation and the intern, knowing how shorthanded they were on the floor, actually wheeled me outside and into the emergency room where they had enough of a staff to give me blood. When I was released on New Year's Day the storm was so overwhelming and widespread that I was barely able to make it to Connecticut before they closed Interstate 84. In fact, even though I considered myself a good driver, it took the driver of a jack-knifed tractor trailer, to help me get around his truck crossing two lanes. I had to shovel my way into my apartment for a good night's sleep. The next morning I decided to go to work even though the roads were horrible. The Milford police stopped me telling me the Merritt Parkway was closed and that GE, of course, was closed as a result. Then they asked to use my four-wheel drive Jeep so emergency workers could get to their jobs. So two days later I did get back to work.

My primary job was writing the Report for Managers and it was this Report that enabled me to actually meet the Governor of Connecticut, Ella T. Grasso, who was the first woman to be elected on her own in the U.S. *(not a wife or widow)*. The Report was chosen by the Connecticut Business and Industrial Association for its annual best business communication award. When I got there to receive it for GE, I had trouble getting in because its male members questioned my being there to receive it. As luck would have it, Grasso was standing nearby when I kept trying to prove I was the editor by giving them a blow-by-blow description of what had been in the winning Report along with my GE identification. I told them how GE jobs related to exports had gone from 6.1% to 11.9% from 1970 until 1974...that Board Chairman Reg Jones said it had cushioned the slump in the U.S. economy. I launched into a discussion of the growing gas turbine business in Schenectady which now had 1800 more employees directly related to export sales. It was then when Grasso, who was the guest speaker that evening, stepped up and invited me to her table to hear more about what I had written. I was to find out that it had been a long road up to the governor's house from her modest beginning as the child of a recent immigrant.

In addition to the Report for Managers, I also took turns with others in corporate communications to produce the GE News Report which went to all employees, over 450,000 then. The edition in 1978 was one I wrote in which I did a fake interview with Board Chairman Jones. We were celebrating the company's 100th birthday *(it was in 1878 that Edison introduced the incandescent light bulb)*. It was titled: "Jones: GE Challenge in the Next 100 Years, Grow and Maintain Our Enterprise." First, I had him note how we were able to shake off several lean years in profits and how each of us can be pleased and proud that last year we were able to make 6.2% in profits on each sales dollar. He discussed how the billion dollars in earnings would be used for shareholders and reinvested into the business to grow sales and jobs. He was pleased to report that $820 million was plowed back in 1976 and thanks to improvements, we expected to increase our investments in the business by 15 to 20%.

The Q&A section had headings of: Profits Invested to Build Jobs, Outlook for 1978, Nuclear Energy Markets, Responding to Change, Economic Uncertainties, International Sales and GE Employees Equal to Next Century's Challenge.

The rest of the issue was devoted to photos and stories headlined: Job-Protecting Investments Help ...at Evendale ...at Louisville ...at Selkirk ...at Pittsfield ...at Tell City ...at Fort Smith and Tyler, too.

Two other pages ran with photos of employees under the headline of "GE Starts Centennial Off Right" with short stories subtitled "Thanks to ...people who have pride in their work ...people with cost-saving ideas ...people who are productive ...people who satisfy customers and people who work safely and people who care (*Jones was also a big promoter of employees helping the communities they live in*). I also ran a full page on seven of GE's early pioneers and at the request of Jones gave him the photos I had used of Steinmetz, Edison, Stanley, Steenstrup, Alexanderson, Coolidge and Sprague.

Anything written for Reg Jones went through Bob Fegley, who wrote speeches for Jones. Fegley was very pleased with the "interview" and the accompanying articles. In fact, he gave me the copy of the note he sent to Reg which I still have today in which Fegley writes, "It stresses the importance of good earnings for reinvestment and job security. She incorporated some ideas and corrections I gave her." Afterward he called me to tell me Reg was very pleased with it. But in the note he had identified me as Marie Kuykendall from CERO (Corporate Employee Relations Operation). I was upset he thought my name was Marie but the Chef told me he would tell Jones what my name was. So when I was over there playing bridge with the chef and his assistants one day, Jones came by and called me Mary.

Another edition I am particularly proud of had the title of "Profits Needed to Build Future Sales and Jobs." In it, I did not make up the interview but actually interviewed Otto Klima, Vice President and General Manager of the Re-Entry and Environmental Systems Division in Philadelphia which was heavily involved with U.S Aerospace Programs. His pitch was technology must be increased to maintain our industrial leadership and strength as a free nation, that technology developments create new and better jobs to replace the old ones. He talked about spin-off businesses from space projects, citing the development of materials which can withstand the intense heat of missile re-entry now finding their way into factories to make more reliable products. Manipulators would aid in designing robotic machines.

In reading this nearly 40-year-old issue today, he is still a man ahead of his time. He predicted that the search for food and energy would require new technology and after that space will be the next job creator. He ended on: "But here again, we need to develop the technology to build the equipment to allow man to do it. The whole point is: we can't stand still. If we do, we lose our good wages and standard of living."

I was glad I still remembered my shorthand from high school when I interviewed Klima. He was a fast-thinking and forceful person. I got an earful when he took a call in the middle of the interview. He started yelling at the caller for having the gall to try to get one of his engineers to give him information on the new laser missile technology they were working on. "All you want that information for," he yelled, "is to endear yourself to the Russians so you can grab some more détente headlines." Before slamming the phone down, he said, "Give a damn about the country that has done a lot for you and don't you ever pull that stunt again." I was stunned when I realized it was Henry Kissinger he had chewed out.

Later, as I was making my way out of his office, a fruit basket was being delivered and Klima reached for it. Klima said Henry must have gotten to Reg about his lack of cooperation because he always sent a basket of fruit before he visited to discuss tough situations. He tossed me an orange out of the basket and told me to be sure to run the interview by him before I printed it which is, of course, what I did.

I also produced a monthly news report for relations professionals throughout the company which I did not enjoy. At the time, I did not fully understand why it was undergoing such intensive review by management with constant changes. But when Ronald Reagan came into the White House, things drastically changed. After the nation's air controllers went on strike, the country's mood concerning union power changed. Reagan, even though he had once been a union man (*Actors Guild*) and later a spokesman for GE Theater, decided to reduce what many considered overreaching by the National Labor Relations Board.

I was given an article for the relations newsletter announcing that Len Maier, GE's Vice President of Corporate Employee Relations, was announcing a new Relations Management Program to train relations people in all aspect of employee relations work throughout the company. A Review Board was also established at headquarters to review the progress and approve written communications. All articles dealing with this new training program carried an asterisk alongside the heading to indicate it could not be reprinted.

When I got a call from one of my union relations friends in Schenectady undergoing the training that he needed to obliterate the fact he graduated from Cornell, which was known for its liberal union relations policies, it certainly confirmed my suspicions that GE was going to take on the unions and the NLRB was going to allow it.

Soon I was running articles about how the NLRB reports "it will quit worrying about the accuracy of what unions and employers say to each other in the heat of union representation campaigns." By a three-to-two vote, the NLRB abandoned a 1962 decision and declared it will 'no longer probe into the truth or falsity' of campaign statements made by the parties prior to an election. "Employees are mature individuals who are capable of recognizing campaign propaganda for what it is and discounting it," declared the NLRB board members.

From then on, the relations newsletter contained articles on plants that were voting "No Union." This went a step further after new relations management training included instruction on the permit procedures to decertify a union. Soon we were publishing how some plants were returning to union-free status. There were no asterisks on these stories.

When votes were close as in the case of one taken at the new Appliance Park East in Columbia, MD, (*1086 for union, 966 against*) the NLRB regional director recommended a new election. Again, no asterisk was on that article. Results in 1976 showed that unions won only 3 out of 15 NLRB elections held. By year end, 41 manufacturing plants completed five or more years without an election.

It was during this time I got a call from a headhunter who said the CEO of United Technologies in Hartford, Ct., was looking for a VP of Employee Relations and that he had been impressed with the Report for Managers. I was stunned at the request because I had a reputation for being anti-management -- especially when I learned that if you were age 35 and over, GE had decided they did not have to abide by EEO rules because it would not be worth their while in management training costs plus there was reason to believe those with that many years of service might be bitter. I had realized this after I was told to redo my personnel files to eliminate the fact I had once had to punch the clock. This message was given to me by a manager at the GE Training Center at Crotonville, who appreciated the fact that I had helped him create a communication program even though I was not eligible to go to Crotonville.

Evidently the headhunter and Al Haig, who headed United Technology at the time, did not know GE considered me too old to promote.

So I went on the interview with Haig. It didn't take me long to get myself eliminated when I noted I was a Democrat. When asked about the Make Schenectady Competitive Campaign, I was sympathetic to those pieceworkers who used their ingenuity to make more money. I also pointed out that factory employees on the same scale needed to be rewarded with incentives to excel with programs like the Suggestion Plan. I was not surprised when he cancelled lunch.

The headhunter then set up an interview with a Consolidated Edison executive who was looking for an annual report editor. We met at the Waldorf in New York City. When I did not play "knees" with him in the cocktail lounge, he lost interest. So did the headhunter. I never heard from him again.

So it was back to the Report for Relations Professionals. While this job was less than satisfying because it required so little writing skill, at least our group was not charged with writing the many benefit booklets corporate published each year for all employees. However, when the Employee Retirement Income Security Act (ERISA) was passed in 1974, I was chosen to get involved producing the booklet explaining retirement benefits. The benefits group writers had spent two months trying to get approval of the draft of a new booklet they had produced to meet these guidelines. ERISA rules by the Department of Labor stipulated that the booklet must be written so it could be understood by an eighth grader. No matter what they added or deleted, it kept getting rejected. When I was chosen out of our group to rewrite it, my first thought was I should never have let them know I was from West Virginia during that United Fund Drive.

When my first draft was rejected, I decided I was going to find the person who rejected it. After several calls to the U.S. Labor department, I finally got referred down to someone involved in the regulations requiring the booklet to be understood by an eighth grader. He told me they had purchased a reading machine to analyze all these company booklets coming in. After many questions, I found out the machine did not tolerate a lot of three syllable words which indicated a vocabulary an eighth grader might not have.

The Labor Department contractor realized as quickly as I did that the word "retirement" was one of these. He could not assure me how long it would take to get that machine fixed. He even agreed it would be best to throw the damn thing away but a higher up had ordered these machines for the department to assure standards were being met. So it was up to me to find other words to substitute for retirement such as "when you go on pension" or "after you leave work and you get money."

The reading machine not only liked the new draft but said it could be understood by a sixth grader. Naturally, I took a lot of teasing for my ability to communicate to all levels. And even the U.S. labor contractor laughed when I told him to run the name of the new Employee Retirement Income Security Act through his machine. Still, it was very frustrating wasting all that time dealing with regulations made with good intentions which led to the road to hell in implementing.

In fact, it was in the middle of this that I got a call from an old boyfriend geologist whom I had met at the YMCA swimming pool when we were learning how to do the Eskimo Role in a kayak. He had left Albany and gotten a job at Shell in their oil exploration division. He was also a pilot. He was calling me from the White Plains airport to say that I had a half hour to meet him there. We could go off for a great life in Canada. Just don't tell anyone about it. Just come to the airport. It was tempting at that time but the not telling anyone about it did bother me. Nearly a year later I ran into an old friend of his in Albany who told me that Jim had died a year ago, that they never found the plane but figured it was in the Gulf of Mexico. I mentioned the phone call and when I got it and that was after his alleged crash in the Gulf. His friend knew him better than I did and concluded that he suspected Jim was running drugs on the side because he was spending money as if he owned an oil well and perhaps he found a way to get out of an oncoming legal situation.

So not being impulsive then had saved me from a life on the run, however interesting it might have been.

Our corporate communication staff also wrote news of corporate activities which was sent to some 150 product department plants, most of which had newsletters. Most of it was promotion of company benefits as well as news of how we must do better than our "voracious" competitors. An abbreviated "Around the Company" column reporting on business practices at that plant, which could be useful at other sites, was something each of us contributed to. The sources for these were newsletters from these plants throughout the U.S. One of our jobs was to "audit" these newsletters to assure they were properly reflecting business needs. We also produced announcements of quarterly earnings reports which always came out on a Friday.

But there was one time when it was anything but routine for me and William, a co-communicator, when we were assigned to get the stock price and earnings news release ready. What we didn't know was that the executive team determining the quarterly earnings report was concerned about giving out a rosy report during a time of intense union negotiations. Security measures had been doubled and the meeting was being held in the front of the auditorium away from the offices. We communicators were often the last to be told because it took time for the news to come down the "need to know" chain of command. We wanted to get out of there before 6 p.m.

Because we both had plans for the weekend, *(he and his family were leaving town for a family reunion)* and I was heading off to a kayak competition in Maine, I suggested that because I knew the projectionist we could get in there and find out earlier what the earnings report was and have the release ready to go immediately. We were well ensconced with the projectionist as we shared our snacks with him when all of a sudden a Senior VP got up from the group below and headed up the aisle to the projection room. The projectionist quickly pointed to his closet-size restroom which we ran into.

William and I spent about two hours in the restroom with William sitting on the john and me positioned sidesaddle above him on the commode top. We could hear the VP say he would be in the projection room the whole time because they wanted to assure themselves that top security was in place because of the sensitive discussions over the effect of an earnings increase in union negotiations. He assured himself all sound systems were off. It was an excruciating two hours with William constantly reminding me he had a family and he hoped I would remember that in the future. Adding to our panic was the possibility the VP might have to go to the bathroom or he would hear one of us cough.

Needless to say, William and his family did not leave on their trip Friday night and I did not get to Maine in time for the race Saturday morning. We were lucky the projectionist gave us the signal to come out right after the VP left after the meeting.

The late 70s were not only a time of civil rights and equal employment opportunity protests but many of the more active protesters were getting out of hand protesting nuclear power. I found this out because I was often in the mailroom checking on the delivery of all the news releases, newsletters and reports we mailed to the plant editors. I was amazed to find that mail addressed to the CEO would contain not just threats, but an occasional packaging of a tarantula between cardboard liners which was picked up by a scanner. "Don't these people know the CEO doesn't open his own mail?" I would ask.

Because of such threats, it was no surprise employees became fearful. It was this fear that caused many to believe an internal notice to all corporate employees that the company had hired an FBI agent to serve as the new security chief. It was said he was known for having kept Bobby Kennedy from entering a federal building because he did not have proper identification. The bogus report noted this chief had acquired two German Shepard dogs -- which every employee needed to be sniffed by -- so the dogs would know to let them through the company entrance. Times for sniffing were announced with those with the surnames starting "A" through "F" to arrive at the security offices at 8 a.m. and "F" through "J" at 9 a.m. and so on. Amazingly, many showed up for this test, resulting in the security chief using the public address system to announce it was a hoax.

The PA system was mostly used for a constant "white noise" which was popular then for promoting a "productive environment." But there was a time when that PA system nearly scared me to death. I had become accustomed to coming in to work an hour early to get the best of a heavily subsidized and lavish breakfast. If too early, I would wait out on the deck for the breakfast to start. It was about 7 a.m. when I looked up to see a hot air balloon with the McDonald's insignia nearing headquarters. As it approached, I could see that the clown, Ronald McDonald, was in it and gesturing wildly. I had read that a new McDonald's was going to open in Bridgeport that day and realized he must be part of it. However, it looked like he was going to have to make an emergency landing.

All of a sudden, the top side of the executive building opened and security officers with guns appeared demanding identification because they feared he could be a nuclear terrorist. Then a demanding voice telling him to "come down now or else" came out over the outside PA system. Ronald gestured wildly at the balloon above him losing air, yelling he was supposed to fly over Bridgeport and could not make it. He was able to land the balloon down in the field below the buildings. He was quickly surrounded by security people at the entrance. It only took them a few minutes to evacuate Ronald and his balloon. It was then I realized what was behind all those walls on the third floor of the executive building contained. I had not believed the CEO needed all that room.

I didn't bother telling my bosses about it when they came in. I had learned by that time to not give out any information about anything I had heard, said or done.

While I got outstanding reports on my job performance *(I always subscribed back then to the idea that women have to work twice as hard to have a man's job)*, my lesson in never saying anything came from a berating I got from the Vice President of Corporate Employee

Relations. I had gotten a call from a good friend at the Research Laboratory in Schenectady who knew the wife of a Brookings Institute man who had been brought in by the company to study pay practices across the company. *(Even today it is still a rigid system of management levels with much lower standard rates for individual white and blue color workers)*. He spent nearly a year accumulating the data. I was entranced by him in an interview. He was recommending employees be paid for performance and responsibility, no matter what their position and level in the company was.

His most vivid example was a corporate secretary who just served as a receptionist. Her biggest mistake might be sending someone to the wrong meeting. Yet, she was being paid three times more than a low-grade secretary in one of the product businesses who could easily make a typing mistake costing GE millions of dollars. His report concluded GE would never be able to compete with IBM in the computer business because GE did not know how to pay individual top performing engineers like IBM did.

To his utter dismay, his report sat on a dusty shelf. The friend I was talking to said his wife had called hysterically telling him that her husband had self-immolated with gasoline that morning. I mentioned it very sadly to my boss who told this to the VP. The next thing I knew I was in his office undergoing the most intense questioning and accusations I had ever heard. He accused me of lying, brought in my manager and both told me never do that again, etc. I was stunned. I pleaded I must have misunderstood. I would tell my caller that it was a lie…that the Brookings man died of a heart attack. Later I was to find out that GE Metropolitan did not pay death benefits in case of suicide and the VP was only trying to help his wife. I was mad that they felt they could not trust me. But then I did realize that was why I was not fired. They knew the truth. I was just glad the family got the benefit because he was the smartest person I ever met at headquarters

After that I kept my mouth shut about anything I had heard. But there were times when I could not prevent them from learning more about me. I was still seeing an old boyfriend in Schenectady where I had worked for 14 years. He was a pilot and we used to fly up to Savage Island in Lake Champlain for the weekend. The guy that owned the island had once been the lieutenant governor of Vermont and had bought the island in the late 50s when he was diagnosed with throat cancer. He went out there to die but instead thrived as a sheep farmer which allowed him to keep an island he had spent all of his money on.

This time I caught a ride on a corporate plane going to Schenectady. Unfortunately, I had gotten into the executive bar on the plane. I was in no condition to buy groceries which I was supposed to. My friend had been flying a helicopter all day doing power line patrol and was dead tired. So we got a late start to go to the island because we had to take the time now to get the groceries I was supposed to get. It was nearly dark when we arrived at the island with its grass strip. The lake was so calm it reflected itself like a mirror. We thought we were landing on the strip when the water started spraying off the propeller. It would have been a perfect landing if we had pontoons.

When the plane was lifted out of the water the next day by a barge which happened to be laying a pipe across the lake at that time, even the eggs in the grocery bags were intact. But the problem for me came when the island was surrounded by drug agents looking for dealers who often flew in from Canada. Thanks to the local sheriff who had been given a ride in a helicopter my friend also owned, we were able to explain who we were and why we were there. But the early reports to the Burlington Free Press had both our names as alleged drug dealers.

As usual, I was back at work on Monday but the Employee Relations Manager at Burlington, Vt., recognizing the unusual name and knowing I was at headquarters promptly faxed the article to the VP of Employee Relations. I was called in and asked how my weekend was when I said, "Oh, the usual." He gave me an incredulous look and handed me a fax of the newspaper article. I tried to explain what really happened, leaving out the part of going up there on the corporate jet. Fortunately, the Burlington employee relations manager called back saying it was all as mistake by the FDA investigators and that a retraction would be in the newspaper.

When I left that job to become a speechwriter in Corporate Operating Services, my immediate boss, Dan Crabtree, did recommend me for it and I believe he was sincere. He even added with a grin that he hoped I would continue to keep my head above water.

MY DREAM COME TRUE JOB

When Reg Jones announced several years before his retirement in 1981 that he was going to start looking for a replacement, one wonders if he knew what a competitive race it would end up being. At that time Jones asked the Executive Manpower Vice President for a list of candidates which totaled 18. From then on there were probably more speechwriters in the company than there were vice presidents.

Having enjoyed my experience writing speeches for the AC Motor and Generator business leaders in Schenectady, I decided what a dream job it would be to write at the corporate level. I quickly got my list of references from engineering and manufacturing technology managers as well as the note from Bob Fegley. I asked Lou Marsh of Corporate Executive Communications for an interview. He put me off saying they had no open positions at the time but I could tell by his disdain that he would delay hiring a woman as long as he could (*and he never did*).

Marsh protested that the place to start was in Advertising and Sales Promotion in Stamford and why didn't I apply there. Because he farmed out speeches not handled by his staff to A&SP, he would recommend a job there for me -- then he would get a chance to see how well I performed. So despite the fact Fegley was his boss *(and I got the feeling he didn't like the idea I had contact with him),* he ignored me.

It was his arrogance that led me to be hired directly by one of the Senior Vice Presidents, Robert Kurtz, who headed Corporate Production and Operating Services. Prior to this job he ran the Industrial Group which included the AC and DC Motor businesses in Schenectady. He was charged by Jones to set up an organization to spread state-of-the art computer capability from the "haves" to the "have nots" across the company to improve productivity. Kurtz was on the list of 18 to be considered for CEO to replace Jones when that time came.

The organization was formed right after Jones had a meeting with the GE Board of Directors in which he pointed out productivity in Japan and Germany -- where wages were similar to ours -- was now outpacing the U.S. U.S. Department of Labor statistics showed the rate of growth in output per man hour in manufacturing in all other major industrialized countries exceeded that of the U.S. -- and that if the U.S. didn't soon match the efforts of it offshore competitors, it would lose its lead in the 80s.

Most of all, Jones cited statistics showing technology investments in plant and equipment alone accounts for 70% of productivity gains. In stating "there is only a limited amount of improvement from the physical efforts of any one employee," he cited studies showing neither the quality of work performed nor a rejection of the 'work ethic' appears to be responsible for the productivity decline. What has played a role in addition to outdated equipment, however, is the diminishing size of the skilled labor pool for manufacturing and a scarcity of people knowledgeable in rapidly advancing technologies."

He cited four major elements present in a highly productive business. 1- Advanced technology. 2. Strong capital investments. 3. Skilled labor ...all of which must be directed by 4- Competent Management.

A "Factory of the Future" program was kicked off by Reg Jones in June 1978. He declared: "Technology investments are the make and break factor in the company's future. We're working hard to make sure the company puts its money where its mouth is in terms of support for technology developments, research and technical manpower. Improving productivity is the way to achieve cost leadership in coming years."

In 1978, Jones called for major investments in new plant and equipment which would result in an annual productivity growth of 6% in the 1980s.

The first group organized by Kurtz was a companywide Computed Aided Design and Computer Aided Manufacturing (CAD/CAM) Council made up of 98 manufacturing and engineering managers representing the company's 43 Strategic Business Units.

Heading up the Production Systems Application Center was Bob Erskine who sent experts into outside companies and research centers to find the most efficient production innovations and practices.

Because of the proliferation of computer systems, many of which did not interface, Jim Duane was named head of Computer Applications System. His team not only assessed and recommended what was available but also developed advanced systems for companywide use.

As a senior vice president, Kurtz had five vice presidents reporting to him, one of whom was Marian Kellogg, VP of Corporate Consulting Services, who had come up through the engineering ranks. She was proof to me the engineering and research types in the company were more open to women and blacks. Because of the technical nature of their jobs, it was less political and easier to determine if the person was doing a good job.

So I asked Kellogg if there were any openings for writing in her group noting that the executive corporate office had no openings at this time. She laughed and said that her boss Kurtz was looking for a writer because he was unhappy with the way he was treated by that office. He was seeking the position as CEO, but was not regarded by the corporate office as a likely winner. Therefore, any speeches he got from that group were at best nearly pointless.

That was enough for me to immediately make myself known to Kurtz. I had an axe to grind, too, having been turned down by Marsh. Kurtz explained he would have to give me a different title because policy dictated that all speeches were to be written, edited and approved through the corporate communication office for major conferences. He would just tell them he was going to do his own to lighten their load. This was fine with them since they did not consider him a contender. Still, all of his speeches would have to be approved by them.

To make sure his speeches were not pointless, I suggested to Paul Quantz, who headed the Computer-Aided Design and Computer-Aided Council that we form a CAD/CAM newsletter and invite engineers to submit articles on technology advances they were making. Having had experience at the plant level, I knew many managers were still used to running their own business under Cordiner's decentralization plan and that they did not appreciate corporate advice.

Quantz quickly agreed and soon we were publishing a newsletter sent to some 5000 engineers. The articles they sent in on redesigning old products and coming up with new ones as well as the use of computers and robots to reduce cycle time were of great use to both of us. I had my examples of how CAD/CAM was working.

If Quantz had an SBU representative who was holding back on what they were doing, he was able to ask the right questions. We had the data from the "haves" to tell the "have nots" about.

For the next three years we held seminars across the country highlighting state-of-the-art technology and how electronically-controlled functions and processes could drastically improve productivity.

To drive the productivity challenge home, we started with 1960 data showing how back then -- when we had modern plants -- that it took seven Japanese workers and two German workers to equal a U.S. worker in terms of output per employee. Then we showed how their investments in new plant and equipment were now paying off. In 1978 Japan installed 7800 robots, more than the total in the U.S. A major GE competitor, Hitachi, had 800 robots compared to 60 in GE.

Thanks to being located in the same area with Quantz, Erskine and Duane as well as the articles sent to our newsletter by front-line engineers, Kurtz speeches were packed with examples of how we were getting back on track in implementing computer technology throughout the company.

One of our major meetings each year was at the General Managers Conference in Boca Raton. It was a highly sought after conference for those wishing to become the next Board Chairman. I kicked off Kurtz's speech calling for productivity improvements by highlighting the word "productivity" on fast-moving slides in foreign languages, which got their attention. After citing data that from 1972 through 1976, GE trailed two companies in Japan, one in Germany and one in the Netherlands in sales per employee, we offered help for massive investments in technology.

The news that in 1978 Siemens had invested $905 million in research and development, much of it in computer technology, was even more unnerving. While GE had spent nearly $1.2 billion, more than half was under government contract which tends to be in the defense business. Siemens also reported that by 1990 they expect 40% of all current manual office work to be replaced by computers.

The GE challenge for the 80s was clear. While GE's productivity of 2 % was better than the U.S. average, it lagged significantly against international competition. Hitachi, Mitsubishi and Phillips averaged 5% and Siemens, with its huge technology investments now, averaged 7.5%.

GE was starting to make major investments but the challenge for the 80s was clear. To meet Jones's goal of 6% productivity growth in coming years, much more would have to be done. Because we had much technology in-house, there should be no proprietary boundaries in linking know-how with need. Our group was charged with providing that link.

One of the quickest ways to improve productivity was through equipment investments to substitute plastic for metal which saved material, energy and labor costs.

The Major Appliance Group in Louisville, which used nearly one-half of GE's total steel purchases, switched to plastic for major components on its top-of-the-line pot scrubber dishwashers. The key to this cost-saving, productivity improvement was a 3000 ton capacity injection molding machine, the world's largest. It molded the inner door and tub using a proprietary substance called Permatuff. Major Appliance also switched from a metal-liner in its 17 cubic foot refrigerator to a one-piece plastic liner. This eliminated fabrication costs, painting and assembly line labors.

Central Air Conditioning's plant at Tyler, Texas shared their know-how of producing permanent mold ferrous castings in-house which eliminated the need to purchase sand cast parts. This was first done on crankshafts for their compressors. Now it was being looked at by Transportation Systems people in Erie, Pennsylvania for iron cylinder liners in diesel locomotives and by Turbine people in Schenectady for stainless nozzle partitions in turbines.

Use of solid state controls was also shared resulting in dramatic improvements in the economics of automation equipment. A copied example was an automated shear line system incorporating the latest microprocessor and programmable controls from the Medium Transformer Department at Rome, GA. This allowed a machine speed increase of 300% thus tripling production of steel laminations and lowering costs by 33% because of reliability.

The Lamp Division plants in Ohio were able to realize and increase production 28% and decrease downtime 10% by changing the conventional stepping switch control to solid state programmable controls.

A new era in metal cutting tools was introduced with examples of how the production of coated carbide inserts could combine the speed capability of ceramics along with the toughness of carbide. We rolled out examples of how this led to tool systems which can increase productivity up to 80% on metal cutting operations.

There were many examples of how computer aided design and manufacturing allowed engineers to dramatically reduce design and manufacturing cycle time. We had videos of several of these including one from Jet Engine in Evendale, Ohio where engineers used video terminals no larger than a TV set to electronically make additions or changes in any segment of the design. It showed how finished designs were stored electronically. When a hard copy was needed, it could be quickly produced by an automatic plotter.

Soon the Mechanical Drive Turbine Department in Fitchburg was using an interactive graphics system to generate cross section layouts for custom designed turbines to reduce cycle time from 20 weeks to 4. Even customer service employees got into the act and used a related program to generate 800 propositions a year versus 400 previously with one third the manpower in one-fifth of the time.

The Gas Turbine Department in Schenectady, which produced 500,000 compressor blades per year, was able to make their investments in computer technology pay off by reducing design and production time from four man-weeks to two days. It took a team effort of toolroom and advanced manufacturing engineers along with interactive graphics personnel to come up with the method to graphically relate the cam to the airfoil contour which defined the cam's analytical shape.

Erskine's group promoted the use of investments outside companies had made to improve productivity. At Caterpillar, computer controlled shuttles transferred work pieces from machine tool to machine tool. At final assembly, computers directed the placement of frames to be moved automatically. CRT terminals showed the lead man how many items to pull from the buckets.

From the ITT Research Institute in Illinois came a hot particulate pressing process for converting aluminum scrap into hot forged parts. Cost comparisons with current manufacturing methods indicated that GE could potentially save over $2 million per user using this type of process.

While CNC (*Computer Numerical Control*) machine investments were becoming somewhat prevalent in many GE factories, particularly in the lamp and power generation businesses, robots were also being developed for machining as well as assembly line work for even greater productivity gains.

One of our big GE successes was holding an annual robot seminar in which we showcased our in-house development of robots. Four of our 43 Strategic Business Units in GE had developed robots by then and were now sharing it with other GE businesses. We also invited robot manufacturers from outside companies in the U.S., Japan and Europe to demonstrate how their systems could improve productivity. Our biggest seminar was held in Schenectady where eleven robots were operating at the Machining Development Operation in one of Edison's original buildings. Seven of these were brought in by vendors on a consignment basis. Demonstrations of the Cincinnati, Autoplace, Tralifa, Unimate,

ASEA, Prab, Scikoshas and Olivetti in addition to three from GE showed how productivity could be improved ranging from tool changing to welding to stacking to painting and assembly.

Duane's group was also working on advanced robot technology which involved voice entry systems to direct robots to perform a variety of functions to eliminate the need for an inspector at each assembly stage. I was asked by Jim to test one of their voice entry systems for the Major Appliance Plant which vividly pointed out the challenge they had in getting the robot to recognize a command. I still had quite a bit of my West Virginia nasal accent. When I told their robot to sort out refrigerator parts coming out of the paint booth and place them on specific conveyors the robot rebelled. It did not like the way I pronounced certain words such as "right, left, push and wash." As Duane said, this just shows the challenge we have because we also have many factory workers who have accents from other countries.

Still, introducing robots into factories, particularly in hazardous operations, was to be a major thrust. One industrial sector identified 170 applications for robots at a projected savings of $4.7 million dollars annually. However, we did caution our audience not to adopt the common practice of giving robots names such as Charlie, George or Mary. This could be insulting to nearby workers. In fact, Kurtz loudly objected when we suggested we have a robot give his next speech.

Most of our presentations were made within GE such as the company's annual Board of Directors, the General Managers Conference in Boca Raton, the Corporate Officers Meeting, the CAD-CAM Council (*every three months*), the Corporate Audit Staff, the Technical Managers Meeting, Crotonville's Management Science Workshop and the Manufacturing Management Conference. There were one-time meetings with the GE Research and Development Center, a Computer Aided Engineering and Manufacturing Control meeting, the Corporate Audit Staff and a

Companywide pre-negotiations meeting before the 1979 contract negotiations in which we stressed technology investments as the best way to be more competitive.

But Kurtz' ambition to be the next CEO made him realize he needed credibility outside of the company as well. One of the prestigious national productivity meetings was the National Autofact Group which met in Detroit to promote automatic factories. We kicked off that conference.

When the U.S. Chamber of Commerce announced they were going to hold a national meeting devoted to manufacturing productivity solutions we were quick to respond and landed a prime time spot.

But the real coup came when Kurtz made contact with a high-level government official who was retiring. This official could get him a speaking spot at the annual U.S. Governor's Conference with the provision he found a good job in the private sector. His reward was a job with our group and about the only thing he did other than introduce Kurtz to all the governors was to set up a meeting with some Japanese government officials who might be able to let one of us in their factories to view their robots.

I could not believe the money we spent on illustrating that speech at the Governor's meeting at Hilton Head in South Carolina. Kurtz might not be able to remember all the investments made in productivity, but he knew a second generation slide when he saw one. To get photos of major GE plants across the U.S., I spent several days on the phone hiring aerial photographers in those cities to get their best shots of those businesses. I already had lots of original slides of their best technological investments from the engineers sending stories to the CAD/CAM newsletter.

However, the day before we were to leave for his speech, I found out about a groundbreaking innovation from an engineer who sent the information along with some photos and slides to the CAD-CAM newsletter. One of the photos was more dramatic than the slide so I made a slide out of it. Kurtz recognized it as second generation at our rehearsal before the speech. He sent me back on the GE Gulfstream jet to get an original even though I protested the slide I made from the photo was better. The pilot got me back in time to get the first generation slide in the tray before the actual presentation. I was disgusted at this waste of money but he just laughed saying I needed to realize the importance of appearance and while I was at it, to get rid of my accent.

As a result of all these trips, I got to know the GE pilots very well. There was one time when all of us applauded Kurtz' action. We were on our way to a meeting in Florida when he wanted to stop off at Augusta, Georgia to see the Masters Golf Tournament. This was no problem for us because the plane was well equipped with refreshments. When we landed at the fixed base operation at the airport, the red carpet was spread out as usual for corporate jets. But this time it was being yanked back up and carried off by the attendants before anyone could get off the plane. The problem was they spotted GE's first black co-pilot. Kurtz himself got on the radio to the tower and ordered everyone to stay on board until the red carpet was returned. Two office employees in suits showed up with the red carpet.

As head of Corporate Production and Operating Service, all of GE's 25 planes were part of his organization. We never had trouble getting to meetings in the U.S. and while company policy did not allow major leaders to fly together, I never had to worry about that.

Much to the concern of Marsh in the Executive Corporate Communications office, we were getting in their way of promoting who they thought would be the next CEO -- Stan Gault, Industrial; Jack Welch, Consumer; Tom Vanderslice, Power; and possibly John Burlingame, International; Ed Hood, Technical and Bob Frederick, who headed up the Strategic Planning Units.

Word was going around that Jones and the Board of Directors were looking for a nontraditional leader. Vanderslice may have shot himself in the foot when he went public with his challenge of being different and how he would change the company. It was the fact he went public that bothered them. GE already had a good reputation. Kurtz had already been ruled out as too traditional. Perhaps my getting him involved in sponsoring a trip of the GE 100 balloon celebrating the company's centennial to headquarters did not help because he was portrayed by the would-be-kingmakers in Marsh's organization as being an old-guard type reveling in the past.

Also the idea for the GE 100 balloon did not come from headquarters which under Jones was celebrating the company's anniversary going back to Edison's development of the incandescent lamp. The GE balloon was the brainchild of friends of mine in Schenectady - Dave Fink, from the GE Research and Development Center and Pilots Dick Weber and Walter Glass. It was quickly requested by many GE plants across the company to be on site for their special motivational programs.

On weekends, I often went with them to these plants celebrating the company's 100th birthday where the festive balloon was tethered. Most memorable was when Glass, who just had a lesson or two in flying it, answered a call from the Saratoga Racing track officials asking him to fly over their noted Travers Race at post time. Rather than turn down the job because the wind determines where the balloon is going -- not the operator -- he agreed to it.

Luckily, he and Weber, who was flying his helicopter with me in it, were close enough on their wind speed and direction estimates to actually be over the track at post time. However, Glass, excited about their success, forgot about his need to look for the parking lot for a landing. Soon he found himself over the Adirondack Mountains. Fortunately he was able to bounce himself off a large Maple tree at the edge of a highland pasture into the field. We were following Glass and was able to tell Dave Fink and his team in the pickup van where to find him. When all of us got there, he was

Rebuilding the GE House Jack Blew Down

having a wonderful dinner with the folks in the farmhouse who had unbelievably been celebrating the retirement of an aunt from GE. Glass was able to add to their celebration and help himself to a fine meal by telling them GE wanted to make it a special day for her.

But back to the search for a GE successor to Jones when there were 18 running for it.

From the beginning, when I submitted Kurtz' speeches to Corporate Executive Communications for approval, it didn't take me long to see that a lot of my facts were taken out and put in other contender speeches in the approval process. So I began leaving out our best facts in submitted copies. When our speech was given, Kurtz would merely say he had added some late thoughts.

My mentor, Kellogg, warned me that this was not good for my career because I was just irritating the corporate office because they knew the Senior Vice President was not capable of knowing all of these facts. "You have to think of yourself in the long run," she advised. She also warned me that our boss was quite a ladies man and when I went to these conferences which were held all over the country, never to accept an invitation to a party from Kurtz.

I told her I was aware of his affairs with two of the secretaries and that he had never hit on me -- maybe because he knew I had been with the EEO task force or because I simply did not attract him since I was always in a ragged rush somewhere. Still, she said, be careful. She added that in his position he could afford any showgirl he wanted, but showgirls had one failing. They were never in awe of him and he needed ego boosting. He had gone through several secretaries who he dropped when they bored him despite their thrill at being with him.

Her analyses proved to be right. One day he told me he had hired a new person, Marcia, to help me handle the conferences.

He met her at the Hilton in London. He had been impressed by her management style and had hired her as his executive conference management coordinator. Her office would be next to mine.

The current secretary he was involved with was furious since she, like the others, had gone to conferences with him to supposedly manage our participation in these companywide meetings. I had not liked most of them because they clearly resented me because I was not partying as they were. They would assign me the worst rooms in the resort hotels where we were staying.

When Marcia came into my office to introduce herself, she made some very valid points as to why we should get along. I would have great lodging at all conferences and there was no reason why I should not be getting the money she was getting. All I had to do was teach her everything I knew about the company. She knew that Kurtz could get bored and she was not going to let that happen. She needed the job. We shook hands. Later on I was to find out she had a wheelchair-bound husband no one knew about. By that time Kurtz was in love with her. He would often question me about her but I evaded personal questions. He was interested in a divorce but Marcia was not. He could not understand it.

When we went to conferences, Marcia and I realized word had gotten around about her special relationship with Kurtz. Still, some of the general managers from the product departments would be astounded at her knowledge of their businesses. Even though I would be sitting at the resort bar with her, a few of them would offer to just buy her a drink at which time she would point out that both of us were thirsty.

Needless to say we became good friends. My salary did improve as she predicted, and I even met her husband who was one of the smartest, well-read people I have ever met. I still have a book, Catch 22, he gave me.

Then at a meeting in Arizona, I made a big mistake. The words from Marian Kellogg of "always find an excuse as to why you can't accept an invitation to his party" rang in my head. What was I thinking when Marcia noted that the presidential suite that our boss always insisted on having was quite an interesting place. She told me about the huge cactus plants that surrounded a moat with a drawbridge that led to a luxurious adobe condo. She had hired a mariachi band to play for everyone. His staff would be there.

I decided to go. After all, he was Marcia's job. When we crossed the bridge, Kurtz immediately poured me a drink from a bottle of 100-year old Ballantine, telling everyone how he had caught me lifting a bottle of it from the company plane. Despite the merry mood of everyone and the lively mariachi music, I still remembered Kellogg's advice. I decided that I would secretly pour my drinks in the many floral planters around the villa. I would need to have all my faculties in this new "anything goes" environment.

I was glad to see that Marcia did not seem unhappy, knowing that she was too smart to have to do this for a living. Kurtz was not that unattractive and, after all, he had told me on one of the company flights he was in love with her.

But he had certainly had enough to drink. He spent quite a bit of time commenting on my naivete. I held back my disgust when he told everyone how I was a union sympathizer and once asked him to use his influence to support two days of sick pay for blue collar workers. He then turned to his finance manager and asked him how much that would cost around the company had he done so.

The finance manager, who had quite a bit to drink himself, slurred out his answer of millllllion....millllllions. It was then that Kurtz, with Marcia on his lap at this point, decided I should get to know his finance manager better so I could understand his cost-benefit analyses. As he lunged toward me, Kurtz saw that I was going to leave so he hit the button to pull up the drawbridge to prevent me from going.

It just took seconds for me to realize I could leap across the moat and get to my room without incident. So with a running start -- and thank heavens I had thought not to have those drinks -- I not only cleared the moat but also a field of thorny cactus bordering it.

But this was not true for the finance manager behind me. I heard his screams when he landed in the cactus patch. But I kept running. I was now starting to think straight. I knew that Kurtz always had a limo waiting in case he wanted to go somewhere. I ran to the driver and told him he needed to go to the villa and take the finance manager to the hospital, with nothing said about it to anyone.

He must have dealt with similar control situations because he immediately left to do it. I got back to my room and spent the rest of the night tossing and turning, wondering what was going to happen and would I be able to find a new job.

Our rehearsal for the technical conference for manufacturing management was the next morning. I dutifully showed up with the projectionist and the script. Everything was ready when Kurtz arrived. I had just placed his script on the dais when he crossed the podium with a stern look on his face. When he saw that I was nervous and obviously not going to say anything but just wait for whatever he was going to do, he just burst out laughing. Marcia then appeared and told me that Kurtz said he had never had a party with so much fun -- that the finance manager was recovering o.k. Kurtz then commended me for my quick action to make sure our little disturbance at the villa was kept quiet. Two weeks later I received a raise with the note that I had performed above and beyond the call of duty in a very sticky situation.

I didn't tell Kellogg about it. I was always grateful when she was also a speaker at our conferences. I always tried to sit beside her at Kurtz' dinners before and after the conferences. It was at one of those Kurtz amused himself seeing our discomfort at his indiscretions. The guy he had hired to get him on the program at the annual Governor's meeting had invited several Japanese officials to the conference on manufacturing productivity to hear how the U.S. was doing in installing robots. At a dinner for them, Kurtz tried to get the officials to allow us to visit their factories. They were dodging the request. When Kurtz noticed that one of the Japanese executives was eyeing me, Kurtz told them what a great speechwriter I was and how I could be of help to them in describing all of their "advances." I was offered a trip to Japan but Kellogg noted that my work here had me much too busy for such a trip and she would send one of her engineers over to share our advances with them.

Kellogg was a true non-political technologist. While she used some of the technology investment examples I had, she mostly did her own writing and was happy with word slides to stress her points.

However, there was one speech for Jules Mirabal, who reported to Kellogg, which I thoroughly enjoyed doing. I knew Mirabal was one of our experts on artificial intelligence and he was a fan of the robots in "A Space Odyssey" and "Star Wars." So I labeled technology as the vital "force" in the fight to improve manufacturing productivity – the force of technology. As he explained how capital investments have the largest impact on productivity investments and increased profits, the slides were filled with graphics made from a set of Star Wars characters I had purchased to illustrate his talk. While discussing competitors like Siemens, we even introduced Herr Vader.

Naturally, in good edifying fun, we described the need for investments to beef up our starships to get manufacturing to Warp 6 -- to boost productivity up to 6% a year as Jones wanted to meet worldwide competition and stay ahead of another persistent villain – inflation.   We ended with a rapid pace of slides showing computers and robots at work. Just as Mirabal ended his speech with the revised line from Star Wars of "May the Force – the Force of Technology be With You," all of a sudden the building started shaking. It was then, and I swear this happened, that we had an earthquake. We were in a turn-of-the-century wooden building in San Diego, the El Dorado, when it hit -- a building that proved as flexible as some of the robots we displayed.

However, despite Kurtz's best effort to get national acclaim, by 1980 Jones had narrowed his list down to five sector heads and the head of Strategic Planning for what he called his airplane interviews. He asked each to present their visions and the programs they would put in place to achieve them. Jones even asked them if the plane crashed and they died, which one of the others would they prefer to become CEO and why. He laughed in an interview with Professor Zahavi, from the State University of New York, in later years that every one of them had replied they would not die but crawl out of the crash.

Years later, some 200,000 employees would wish that Stan Gault, who had believed in technology as the key to future growth, earnings and job security, would have become CEO.

However, Jack Welch had impressed both Jones and the Board of Directors with his ability to produce record earnings at the Major Appliance business despite heavy competition. What they did not know, until it was too late, was that he did it by laying off hundreds of engineers well before year end and doing away with the product development laboratory there. This would become evident to them in 1981 when GE's entire line of refrigerator compressors had to be recalled.

But it was too late. Welch was in the CEO chair. In the next 20 years, Welch would do an 180 from Reg Jones's program of increasing productivity by investing in modern equipment and processes. To transform GE into a financial company, Welch would sell off product businesses and advanced technology to the point where 85 percent of GE revenues came from financial services and just 15% from what was left of products. Basic research would be stopped as well as product development labs across the company. The company would no longer set aside 10% of its profits for research and development. Our CAD/CAM organization was one of the first to go.

Before it did, I had managed to get three straight weeks off to work on the 1980 Olympic at Lake Placid. I found things to be high-handed there, too. Many of my old Schenectady Wintersports Club friends led by Dick Weber were able to work the alpine events by training at a World Cup and Canadian-American race there. I ended up on radio control at the start of the men's downhill because so many IOC (*International Olympic Committee*) officials were listening in on our battery-supplied communication system. It was said you could hear me without the walkie-talkies. It was during the training runs to seed competitors that I became disgusted. The Austrians were the most competitive and once when they saw one of their team members not getting a good start they had someone run across the course so he would get another start. We stopped that.

There was also the time when ABC was filming the seeding run and they were waiting to see how a young racer, Leonard Stock, who had been groomed from the age of six to be a downhill racer would do. He was judged to have slow twitch muscles, versus fast twitch, which would be better for downhill rather than the slalom. Because he was new on the racing scene, he was low in the seeding. The ABC crew was impatient waiting because many of the competitors from countries which do not have ski areas were slow. They demanded that our starter start putting three racers on the course at one time to speed things up. When a racer fell in the direct line of the downhill racers, ski patrol ordered me to hold the next racer so they would have time to get the skier off the course. The starter was going to obey but the IOC official there ordered the race to continue. All the ski patrol could do when I radioed back the next skier was on the course, was go out and pull the racer out of the course even though he was yelling about his back.

But the Olympics were a lot of fun. Two of my best friends, Cash Jones and Paul Lozier, were on the maintenance team. Part of the job was to spread pine boughs along the downhill course so the racers could see the boundaries, particularly when they were going at speeds around 70 mph. At one point a falling skier had taken out a whole row of the bough markers. I immediately told the IOC official that Cash and Paul, who were at the top near some bough bags that they would be perfect choices to send down the icy downhill trail to do a quick spreading of boughs. Fortunately, they made it down the course without spinning off the ice and did their job. But they gave me a rough time later when they realized what could have happened to them.

Then there was the time after a warm day of a seeding run when the Austrian coach asked me *(there were nearly 1500 of us in blue suits working all the venues to signify we were workers)* to take their jackets down to the finish line. I had to face the icy course myself. I decided to put on their jackets with the last one being the coach's because he was a large man. I fell on the steepest pitch and when cartwheeling down it to finally land under a liftline to be seen by all. The next morning before the actual race, the coach berated me loudly in front of the competitors. It seems everyone thought he had fallen. The Canadian team became my favorite after that because they just stood there and laughed even though we were all supposed to be quiet at the start so the competitors could concentrate on the course.

We blue goons, as the competitors sometimes call us, even became a little competitive ourselves. Because the Americans had unbelievably made their way into the ice hockey finals against the Russians, the event was sold out. Some ticket holders were selling theirs for hundreds of dollars. I had a bet with Bill Kornrumpf that I could get in with my blue suit and he couldn't. He figured the "maintenance" label on his suit would get him in if he followed the Zamboni ice resurfacer in the arena. But he was caught. In the meantime, because I had a temperature of about 105 from all the foreign flues I was picking up, I felt faint at the entrance and was taken into the first-aid center. I was able to filch a nurse's uniform which I put over my goon suit and got into the arena. Once there I ditched the uniform, and headed for a seat when an official, seeing the word "control" on the armband, ordered me to clear out a crowd that was sitting in the aisles for a better view. I gladly did as told. I even had an unobstructed view of the game. In fact, Bill was among the workers who had gone to the Holiday Inn to see the game on Canadian TV. They spotted me immediately in the empty aisle. The one thing I will never forget about the game was the confused look on the faces of the Russian competitors when the U.S. spectators were cheering even when there was no action on the rink. I think that won the game.

I should also add there was confusion among the officials and companies who had to quickly put together the facilities for the Olympics as well as housing for the blue goon volunteers. For example, Ford donated a lot of station wagons and they not only looked alike but many of them used the same key. So we had to be careful to make sure which one was ours. Many of us were in trailer parks which had been quickly set up. Our trailer had a special feature. The hot water did not come on in the wash basin. Instead it was in the toilet bowl.

Anyhow, I digress. Back to GE and the beginning of the Welch transformation of GE into a financial company. Stan Gault went on to head Rubbermaid where he proved that a U.S. commodity business could be very successful with investments in new technology. As for Kurtz, he took his golden parachute. His staff, including Kellogg, soon followed him in retirement. As for Marcia, she had correctly predicted that she would not be welcome in other jobs in the company. But now she had a resume and a wonderful recommendation from Kurtz. She landed a job planning meetings for the utility industry.

Speechwriters galore were no longer needed. Welch also got rid of the huge Advertising and Sales Promotion Department and farmed this work out to agencies which also wrote books extolling his business acumen. I came back to my roots in Schenectady and worked as a freelance writer for engineers while going after a Master's degree. I also rejoined GE in Gas Turbine Marketing, a new field for me.

What I was to see in the next 20 years was the move from paternalism and a sense of corporate responsibility under Jones to stark materialism under Welch in which a new class of GE workers were added to the traditional blue, pink and white collar. We called them Gold Collars. They were mostly MBAs who had no pride in products or technology but felt great satisfaction in making money on money, especially when much of it landed in their pockets with Jack bagging the lion's share.

THE HOUSES JACK HARVESTED AND BLEW DOWN

When Jack Welch replaced Reg Jones in April 1981, the bold new leadership he called for was immediately felt at headquarters. Welch was, as many put it, 180 degrees from Reg Jones. It was not just the four letter words he used to "distinguish" himself quickly as a force to deal with but his actions. The first thing he did to gain control was undo the Jones structure of three sector heads, 10 groups, 45 divisions and 190 departments.

Long range planning was a thing of the past now. He got rid of the SBU (Strategic Business Units) which were involved in the future planning process central to the company's success in managing such a highly diversified business.

A three circle strategy was substituted with one circle titled "Services," another "Technology" and the third, "Core" with six businesses listed outside the circles. This was accompanied by an edict to all – "Become Number One or Two in your business. Fix, Sell or Close!" If you were outside the circle your days were numbered.

Outside the three circles were Large Transformers, TV and Audio, Mobile Radio, Small Appliances, Switchgear, Wire and Cable, TV Stations, Ladd Petroleum and Microelectronics. The worst part about being outside was the fact your competitors knew it, too. Competitors of many of these businesses quickly lowered their prices to hasten the process. The situation is not unlike what our troops faced in Iraq and Afghanistan when withdrawal dates were announced. In this case, the terrorists could also make inroads or bide their time.

In the restructuring process of getting rid of a couple layers of bureaucracy, which was at first good news for those of us down the food chain, Welch became totally powerful. He was the "decider" as former President, George W. Bush, once called himself. By eliminating the sector and group heads, Welch had free reign to do what he called making GE into a lean and agile company.

His platform was at the Company's Management Training Center in Crotonville, New York where he could select and groom the managers who followed his vision. It was there he created the new Gold Collar class.

A reorganized Human Resources staff became an important tool for him as he gave them greater hiring and firing powers. He also hired a huge legal, public and government relations staff to do his bidding as well as create loopholes to avoid taxes. The famed GE auditor team under Jones was told it was time they became corporate supporters, not corporate policemen. Finance managers were told to start making money, instead of counting it. They too began working with lobbyists to obtain tax breaks.

Over the next five years, Welch would personally meet with more than 18,000 high-potential managers at Crotonville, all of whom knew their career depended on Welch. When he conducted his "presentations," the involved and committed Human Resource person was in the back of the room taking notes. There was no air cover -- no sector or group executive to provide backup for anyone who disagreed with Jack's financial vision.

To assure all GE managers and employees got the message and followed his vision, a new job evaluation system was put in place in which business managers were to list the top 20, the middle 70 and the bottom 10 in their organizations. If there were 20 people on the management staff, he wanted to know the four in the top 20 and the two in the bottom 10 by name, position and compensation. Just as calls for a 10% across the board layoffs to meet his earnings projections to security analyst were made, this process also encouraged shrewd managers to have scapegoats ready to lay off. Good managers who ran a tight ship were penalized. They soon learned they also had to be "creative" to keep their good people in their fight for survival.

Weeding out managers who did not follow his vision of becoming a financial services company was done through a program of what Welch called the "Four E's of GE Leadership," involving high energy, ability to energize, edge and execute. Managers were graded as A, B, and C players. Session C Sessions were conducted by Human Resources at every plant to grade the business and its people.

Incredible amounts of time and money had to be spent on putting together Session C binders full of information to justify their business. It became a joke that Welch paid more attention to the photographs of the managers in the binders than he did the content. This resulted in much emphasis on getting new photos of managers to look aggressive and serious. Because Welch (*he became CEO at age 45*) had been quoted as saying everyone is obsolete after 50, much emphasis was on adding hair and eliminating fat. This was costly because we did not have Adobe Photoshop then. The photos had to be airbrushed and reshot. A Session II conference was held two months later at each plant to see that his changes had been made after the Session I reports were reviewed.

The old Corporate Management Development forms submitted by employees introduced and used by Jones had also been changed. No longer included was the page asking for a list of community projects in which you were involved. It was replaced with a page rating your passion for business. In fact, management soon learned that it was best not to have your residence in the city where your plant was relocated because getting too involved with the community could lead to commitments. Moving to high-end developments away from local politics was best to avoid contact with local officials, especially when the business was in a fix, sell or close mode.

It was an ironic change. Jones had been recognized by government officials for his sense of corporate responsibility and integrity when he served on the U.S. Council of Business. When he celebrated the company's 100th anniversary in 1978, every town and city with a GE plant was involved in the celebration.

But all that changed with Welch as he dealt with the Reagan, Bush and Clinton administrations. Lobbyists would now be the spokesmen for GE. Soon "Spheres of Influence" was the code word for private/public collaboration. A cozy partnership was formed in which government officials *(mostly with campaign contributions)* bought into the trickledown theory promoted to the public of more money at the top creating wealth for all. The problem is most of the loopholes, big enough to what one pundit called to drive a truck through, were handouts without promises in return.

This Spheres of Influence policy not only enabled big business to find cheap labor offshore, but it was helpful to Washington politicians in foreign policy to know that big business could help them deal with developing nations to assure their cooperation. Business wise, the loopholes became so extensive that incredibly today tax breaks are still given to move plants offshore as part of the spheres of influence private-public collaboration.

This support, which some called corporate welfare, was not just at the federal level. Politicians at the state level began giving GE and other major companies tax breaks and cash incentives to move or build a plant in their area. This set off another round of tax breaks and incentives from the county politicians in these states for location there and sometimes within the county's townships.

The Welch team was so good at taking advantage of these loopholes that for the first time in its history, the company did not pay any federal income tax the first three years.

But I digress. It was the houses that Jack blew down that earned him the title of "Neutron Jack."

Just between 1981 and 1990, Welch noted in his book, "Straight from the Gut," he freed up $11 billion of capital by selling off more than 200 businesses in the U.S. In that time, Welch bragged he had made over 370 acquisitions, investing more than $21 billion in such major purchases as Employer Reinsurance, Kidder Peabody and CGR Thomson, the French medical imaging company. The bulk of the acquisitions involved financial services. But GE Capital also bought businesses to harvest including a greeting card company and a used car exchange.

When Welch retired in late 2001, his famous three circle strategy looked like a planet re-alignment. What survived in the Core circle went into the Technology circle where much was also harvested and sold. His replacement would have to deal with a failing Saturn size Financial Circle trailed by struggling Plutos.

You can check out books like Tom O'Boyle's "At Any Cost, GE, Jack Welch and the Pursuit of Profits" as well as two of Jack's books, "Winning," and "Straight from the Gut," to find out the diversity of businesses he harvested and sold in the U.S. – all of which resulted in a loss of some 180,000 U.S. jobs and competitive technology for future jobs.

Yes, it is amazing how easily he turned GE into a financial company. Rarely did the Board of Director's question his actions. As

Welch notes the Board of Directors were also great friends to him. One might add that if they were not, they would soon be leaving as did the head of Cornell University who did not go along to get along. *(From my experience in being in the projection room at a Board of Director's meeting when we were asking for $30 million for CAD/CAM projects, there were no questions but a comment about the hors d'ouevres. Back then they were getting $50,000 for each quarterly meeting and would receive a GE pension for their efforts. Welch doubled the benefits adding stock options.)*

Still, even with their backing, one has to ask how he could dismantle and harvest a company as huge as GE as quick as he did. I was to learn how when I left headquarters and came back to the Schenectady plant. One of my jobs in marketing communication was writing for the head of gas turbine engineering in Schenectady, John Patterson.

Patterson was in trouble for not meeting the Welch schedule of laying off 20 percent of his engineers each year. Dennis Donovan, Welch's new Vice President of Human Resources, had trouble getting rid of Patterson for insubordination in resisting layoffs. Patterson was very popular with the design, manufacturing and service engineers and this could be a problem. They still needed to keep the business going if they were to successfully harvest and then sell it. Patterson was also very popular with CEOs from industrial and power plants internationally as well as in the U.S. He would have to be handled delicately.

The Human Resources VP was able to get rid of George Cox, who headed up the Large Steam Turbine Business in Schenectady for not passing his Session C directives to downsize. Fewer customers (*mostly U.S. utility heads*) were involved so they were not as concerned in replacing him. Moreover, Welch had already condemned large steam turbines as a commodity business because he felt it had already achieved a lot of its technological capabilities – therefore it was defined early as a business to harvest.

However, Welch's new breed of Gold Collar management, consisting mostly of MBAs, still had to contend with the old guard steam turbine marketing, manufacturing and engineering heads who believed there would always be a market for steam turbines and generators. They pointed out the steam turbine business was still making a profit because it had the lead in technology.

Despite the investments that had been made during the Make Schenectady Competitive campaign along with investments in automating under Jones which were starting to pay off, Jack would have none of it. Faster money could be made by harvesting the business and then selling it.

The first series of buildings in the Schenectady plant ordered to be taken down by Welch were the steel foundries. They had been modernized just five years earlier with a new casting process. However, outsourcing was showing it was cheaper to get steel forgings and components from Japan. So the wrecking ball hit the foundries -- which also got them off the local tax rolls - even though the costs of tearing them down were substantial.

I remember being down there with the manager of manufacturing when the Bldg. 95 steel foundry was being knocked down. He started out as a section manager in that business and was speculating that as soon as our offshore suppliers found out we did not have our own in-house capabilities anymore, they would raise prices. It had already happened when the wire mill, which produced copper wiring for the motors and generators, had been closed. Prices had gone up. I guess he trusted me not to report his lack of vision because he went on to note that it would take years to make up the costs of tearing the building down. He lamented that the city had even offered to reduce the tax rates on some buildings not being used just to keep them up. But they, too, were being ignored.

The AC Motor and Generator business was moved out in 1986. Some sections went to Mexico and some to Brazil. The hydro generator business went to Canada and small motors went to Fort Wayne, Indiana which later went to Mexico.

Managers in the motor businesses, who had desperately tried to save these businesses, knew early on they were about to be put outside the three circles because Welch also considered them a commodity to be harvested and sold. The manager of marketing, Vern Mize, took the unusual step of trying to get Welch to change his mind by putting together a sample ad to advertise the new motor they had just introduced to the market. He had heard how Welch had commented on the effectiveness of the Chrysler ad with CEO Lee Iaccoco touting new models they have developed to save Chrysler and to get government support. The ad he sent to headquarters for Welch to approve had a photo of Welch noting GE motors had again raised the bar in performance and reliability. But Welch had already updated his three circles, putting motors outside of it so the ad was never approved. Ironically, the motor line was named the Phoenix.

The sale of the entire aerospace businesses by Welch to Martin Marietta for $3 billion was questioned by some security analysts as dubious because defense business was forecast to grow at an even greater pace in the U.S. because our government was taking more active global roles as peacekeeper by involving itself in wars.

Among these businesses going to Martin-Marietta was Schenectady's Knolls Atomic Power Laboratory which built nuclear submarine reactors. This affected other businesses not in Aerospace such as in the Schenectady plant's power tubes for radar and space. With GE no longer a key player in military markets, these businesses soon disappeared.

Welch's rule was to make no less than 15 cents on the dollar or you are out. The defense business was a fixed cost of 12 cents. Welch also did not like the heat the press was giving him over nuclear pollution at the Knolls Atomic Power Laboratory.

Rumors were flying that the large steam turbine business was going to be sold to Siemens and the gas turbine business would be moved.

In just the Schenectady area, GE employment dropped by over 10,000 in five years and another 15,000 when the large steam business ceased to exist and gas turbine manufacturing was moved.

Nothing could be done to stop Welch and his Gold Collar team. Sick humor among employees may have begun at the Major Appliance plant in Louisville, Kentucky, when Welch placed the Central Air Conditioning outside the circle to sell or close. It was the Louisville plant Welch had gutted by getting rid of the product development laboratory and firing enough engineering and manufacturing people in the second quarter to bring in impressive year-end earnings. At that time, the Board of Directors and security analysts were not evaluating the results in terms of losing a competitive edge for future earnings and growth. Welch was propelled into CEO chair for his prowess in increasing earnings as measured by sales per employee for that year -- which was much greater than that of other consumer manufacturers.

Still to be fair, during his first two years in office Welch did continue Reg Jones' program to invest $500 million to modernize the Appliance Park in Louisville as part of the Factory of the Future plan. He also continued to invest $135 million to modernize the locomotive plant in Erie Pa and $103 million in magnetic resonance imaging business. But he did not continue any of Jones' planned Factory of the Future investments in Schenectady, Pittsfield or Fitchburg.

Welch also followed up on Reg Jones' plan to build a new two story building at the R&D Center dedicated to electronics and computers science at a cost of $130 million. But then within two years, he did a 180. From then on it was fix, sell or close -- no more major investments in plant and equipment, product development labs or basic research unless the government paid for it.

Some pundits say it was the Major Appliance investment that caused Welch to toss the Factory of the Future plan to the wind. The fix no longer seemed an option. His idea of fix often caused major problems such as his insistence that a rotary compressor should be used in refrigerators because of the huge savings of having fewer moving parts. Parts could be turned out every six seconds, an amazing feat given the machining tolerances needed in the new Columbia, Tennessee plant built by Jones. However, the problem was many of the veteran engineers, who knew it had been tried before, had been laid off by Welch. For those that were still there he set impossible deadlines and canceled much of the usual testing before it was introduced to the market. The recall cost $450 million. The word fix was no longer in his vocabulary after that fiasco. It was now harvest and sell or move to cheaper offshore labor. His major drive now was to turn GE into a financial company to make huge profits fast by buying, harvesting and selling businesses here as well as offshore in addition to providing financial services. Thus, he began touting the company as a global player.

Welch's disdain for the Major Appliance Business in Louisville and his message that GE was a global company now thanks to him, was quite clear when a new engineering graduate asked Welch at the management training center that if his performance was highly-rated at the end of two years, could he expect to be put on the high-potential list for manager of the Major Appliance business in Louisville. Welch, who was especially irritated at the geography limits the young engineer had -- especially right after his talk on going global -- advised the young man that if he liked Louisville so much, he should get a job with the (*expletive*) post office.

While this was going on, security analysts were hailing Jack for his financial know-how in meeting profit targets. Others used the word "shrewdness." But it didn't matter. Jack never missed an earnings prediction, even if it meant a quick sale of advanced technology or a layoff each quarter to get it. The business press was hailing Jack Welch as the most admired CEO of the world. Stock prices were rising.

As for me, I was absorbing a few more jobs when our gas turbine marketing group was merged with that of the Steam Turbine Department. This caused even more rumors the Gas Turbine Department would be moved to South Carolina with some of it going to Nuovo Pignone in Italy. Then the Wall Street Journal carried another story of Siemen's interest in buying the Large Steam Turbine Business in Schenectady.

Much of the engineering and manufacturing management had decided there was nothing they could do to change Welch's mind about modernizing plants, particularly those in New York State. Even though Schenectady GE's tax rate assessment was lower than Mike's hot dog stand and the local utility gave GE a lower electrical rate because of high volume, there was at least one state official who made it well known what Jack felt about the state.

New York State's Majority Leader of the Senate Joe Bruno, who had an ego as big as Welch's, publicly acknowledged that Welch was irrational in a meeting with the governor in which Welch said he would do everything he could to get rid of every job in New York.

There were rumors that Welch particularly disliked New York State because he resented the fact that when twin-engine helicopters were made available to the civilian market by Sikorsky, the first one went to New York Governor Carey rather than to Welch.

Another opinion was published in a letter to the Schenectady Gazette. While there was no paperwork evidence of it *(Schenectady GE's employment records had been purged during the EEO years because of possible discrimination evidence)*, Sheldon Steuer, a retired accountant, recalled that as a young engineer, Jack did not cut it in an interview in Schenectady and was rejected for a job. Writing that Welch then got an MBA, he noted that Jack "was a man of great pride, and even greater vindictiveness, and got revenge on those early 'enemies."

He went on to note: "The Schenectady plant, one of the most successful and most profitable of all the GE facilities involved in heavy manufacturing, has through Welch's reign, been starved of modern equipment and pretty much condemned to operate with obsolescent hardware. Nevertheless, by sheer ingenuity and technical competence, plant personnel have met or exceeded their cost and production goals. This, in spite of the constant sniping and rumormongering alleging that the plant was losing money. That this is false is now a matter of record."

While I was not privy to financial records like Steuer was, I remember when the turbine marketing manager, celebrating one of his best international sales, figured he had made a 30% profit on a Saudi Arabia deal. He attributed it to a deal he had made with a privately owned west coast engineering firm. Because GE was a publicly-held company, it could not meet the sheiks request for a Tiffany statue of the Egyptian sphinx which would cost nearly a million dollars. So GE became the subcontractor to the privately held company which was able to add the cost of the perk to the job. It was a lucrative deal for both companies.

As he and others often reminisced, the days when a profit of 5 to 7% was considered good were over. Welch was now demanding 15 cents on the dollar and he was going to get it one way or the other or as he liked to call it, becoming more lean and agile.

I must say that in the first two years of being lean and agile before the massive onslaught of Gold Collars coming to Schenectady, I was pleased to see a lot of the bureaucracy gone. I did not mind absorbing extra work. There were many of us who appreciated not having to wait on approvals to get things done. In fact, even my direct boss, the marketing communication manager I reported to, didn't try to manage me. He had gone into the survival mode of just saying the right things and biding time until retirement.

Therefore, I was free to work directly with the engineering and manufacturing managers in their attempts to assure our utility and industrial companies we were not going to abandon them. We continued to promote our advanced technology lead and manufacturing know-how at state-of-the-art seminars and any U.S. meeting where we could reach them.

At this point, there was still money in engineering budgets so despite the fact that money had been taken out of marketing budgets for brochures and videos, Patterson came through with funds to produce what I considered the best brochures ever produced. Because so many plant managers were afraid to sign anything for fear of reprisal from Welch, it would only take a month including printing time to get a brochure out. We did not bother them for approvals, a process that in the old bureaucratic days could take up to a year as changes were made on changes. The engineers wrote most of the copy and took most of the pictures of their projects. I just edited it to fit and laid it out for the printer. As a result, the brochures were not only timely but also produced at half the cost.

As for the "sheer ingenuity of engineers" spoken of by Steuer in his letter to the Gazette *(I did not know him, he wrote the letter when he was safely retired)* I must say that Patterson epitomized them all for determination and dedication. He took a top gas turbine customer hunting once in Vermont. They came upon a gas turbine operating at a small hydrogenation plant. To prove his point about the reliability and durability of the turbine, he shot into the side of it to have the bullet ricochet off it. The Con Ed customer was so impressed he bought a whole fleet of them and put them on barges outside of New York City for backup power.

Instead of spending money on multi-slide projections, engineering took what money they had been allotted and put it all into producing a top-notch video at a customer site and using that as our blockbuster for technology meetings.

I was again on the road *(albeit commercial flights)* to meet video crews to tape customers who liked their GE turbines and were quite willing to show them off at their facilities. One trip I remember most was in Texas where we were going to an Occidental plant in Corpus Christi. The film crew of two and I were to meet the customer representative at a roadside park. He would then escort us into the plant. When we got to the site, he was not there so we waited. Soon, a guy covered with oil showed up in the parking lot. He told us he had gotten there early and had taken a walk when the ground caved in under him. He managed to get out of the hole which had been caused by an oil pipe leak. He needed to get home but called ahead to see that we got in the plant.

The plant manager met us, glad to hear his engineer was o.k. He gave us free reign of the plant. I almost got us in a similar dilemma as our engineer host had in taking his walk in my desire to get the best shots possible. I had spotted a tower where we could shoot some aerial scenes *(we had no money for the real thing)* when all of a sudden we found ourselves covered in hot steam. We were able to get down before we were fried. But we had some good footage. I was enjoying being competitive so much that when an edict came down asking

Schenectady employees to take a week off without pay to meet Welch's demand for more earnings, I went for it. But I have to say it was mostly because I was so tired I needed a break.

Life without bureaucracy for two years was also made easier for engineers who did not have to work with assigned marketing or public relations personnel to illustrate their talks with photos which were time-consuming and costly. We used the viewgraph machine. It is true that some engineers would almost put their entire talk on viewgraphs *(used before computers and power point)* which made it difficult to read from the back rows. But as Patterson said, utility executives are almost always engineers who came up through the ranks so they are not going to think less of us. In fact, they may appreciate the cost savings themselves, especially when they were passed on in lower turbine prices.

Adding to Welch's dislike for New York was the edict to clean up PCB pollution contaminating the Hudson from the Waterford and Hudson Falls, NY plants. These plants built capacitors and transformers using this fire retardant chemical which was banned in 1972. The government ordered the river to be cleaned up and placed a ban on eating fish, causing a downstream loss of a $6 million dollar fishing industry.

Many felt Welch spent more fighting the charges by hiring a slew of lawyers and a huge staff of PR people (*including some from the EPA itself*) than it would have taken to clean up the river. Millions was also spent on hiring independent labs to question the cancer causing charges. There were so many ads on TV back then protesting the EPA ruling one is reminded of all the political Super Pac ads today telling us how awful each Presidential candidate is. So when Jack harvested and sold off the power distribution businesses, no one was surprised.

The pressure to increase earnings was intense on management. Some said that this pressure, along with the personal desire for stock options and bonuses, caused many to cross the line. The pressure became apparent when the Wall Street Journal, which back then had a raft of investigative reporters who followed the paper's policy then of integrity and stewardship as a corporate responsibility, started questioning some of the deals. The first indiscretion revealed by the Wall Street Journal was about the price fixing of time cards in the Philadelphia switchgear plant. One of the biggest fines GE paid was $69 million when an aircraft engine executive conspired with an Air Force General to divert money from the sale of aircraft engines for Israeli F-16 warplanes. Then there was the charge of a price fixing scandal in the artificial diamond business with DeBeers which was eventually dropped.

But no one seemed to question Welch's decision to turn the company into a financial one by selling the advanced technology and then the business itself. One of Welch's first sales was Calma, a computer-aided design company, which Jones had purchased for $110 million in late 1980 before he left as part of his "Factory of the Future" plan. Welch sold parts of it until it no longer existed in 1989.

Welch also bought RCA for $6.3 billion and combined its consumer business including TV with that of GE. This leveraged buyout proved fatal to RCA. Welch sold off all of its assets except NBC. Some 50,000 employees lost their jobs and their famous Sarnoff Research Center was closed. While a lot of money was made selling off RCA businesses, Welch soon lost a lot of it. Welch used $3 billion he got from the sale of what was left in the Consumer Electronics business to do a deal with France's Cie Generale de Radiologie (CGR)-Thomson to expand GE's growing medical business. Unknown to him, the company was in financial trouble and facing heavy lawsuits from faulty x-rays. GE had to pour $300 million into it to get it back on its feet.

One of the toughest things to deal with was not knowing what Welch was going to do next and in some cases, it would have behooved him to have at least consulted those affected.

John Harnden, a key manager at the Electronics Laboratory at Schenectady's Research and Development Center (*which ironically had been dedicated by Welch after it was planned and financed by Jones*), was in California at a U.S. manufacturer's meeting for customers to announce the digital revolution of GE appliances which would give GE the competitive edge for years to come. As his team was explaining the technological advances in the consumer businesses, an attendee came in with the newspaper announcement that Jack had just sold the small appliance business.

Welch's rule of "If you are not No. 1 or 2 in your business, you are out" was felt loud and clear. It was true that the small appliance business was highly competitive and GE had lost some of its market share. But it was too late for John and his team to show how GE could be a leader again with advanced technology which had been developed under the Jones regime of returning 10% of profits into research and development.

While severe austerity programs were being introduced in what was still left of the manufacturing plants in the U.S., Welch was not only getting obscene pay and stock options but he was also giving huge salaries and stock options to the Gold Collars who supported him.

As a company known for its white collar, blue collar and pink collars *(what secretaries were called in the old days)* it was said, particularly by technical people, that this new class of Gold Collars, in their harvesting of the business for profits, would kill the business -- that their insatiable greed would end up leaving the most diversified manufacturing company in the world in tatters.

Where did the money come from to transform GE into the Financial Services business on such a large scale? The engineers will tell you most of it was from the product departments as well as investors, some of whom were unaware of how the profits were being generated but were impressed with earnings reports. Then there were those insiders who knew about the harvesting and knew when to sell.

Every stone was turned over to get money out of the businesses. Those plants that did survive would see a 20-year drought in new plant and equipment. No longer did GE win nearly a third of the world's industrial research awards. No longer was 10% of GE profits going to basic research. Moreover, product development laboratories across the company were closed and thousands of technologists were laid off.

Welch hired what he thought were top financiers from Wall Street to acquire and manage many investment and financial businesses. They were able to make even more money at GE with pay levels never seen in the industrial, consumer, defense, power generation, and electronic businesses. Wages at the top were no longer 15 times greater than the average engineer. It was 200 and 300 times more, with Welch getting 500 times more in salary and stock options.

The biggest affront of all to the manufacturing and engineering force in the company was when GE Capital took over the financial end of selling products such as turbines and generators on a global scale, particularly when it involved working with the World Bank, the IMF, other quasi-government or even the government itself in third world countries receiving aid. Often sales were politically made under the private/public "Spheres of Influence" policy. Therefore, GE Capital employees were the ones credited with the sale which managers would later call part of Welch's "financial engineering."

There were even times when such deals resulted in a below cost of production sale to satisfy a foreign policy -- which had to be made up by a domestic sale price increase. This did not bother the Gold Collar MBAs in GE Capital, who still got healthy stock option increases through financial engineering.

In fact, Jack's stock option plan resulted in payouts to managers who followed his vision going from $6 million in 1981 to $52 million in 1985 while his own salary and options headed to the top 10 Fortune 500 compensation list. *(I could not find figures for how much he gave out in Incentive Compensation and stock options in the 90s.)*

But the loss of jobs and its effect on the communities when Welch and his Gold Collars moved businesses and jobs offshore was devastating. Back in the Jones era many blue collars workers had feared they would lose their jobs to automation. But Welch, as a short term thinker, knew a quicker way of making money was to put the plants where wages were measured in cents per hour, not dollars. I remember Kurt Vonnegut's first book, Player Piano, *(yes, he started out in GE)* in which he predicted robots would replace workers. He had not met Jack Welch who found even cheaper labor offshore. The irony was that in some cases, skilled blue collar workers were asked to go out of the country to train the new workforce.

I remember an assembler I met at Scarlata's bar in Schenectady who had been to Mexico to train the workers on the motors he used to build. He came back to find his severance pay waiting for him. While in Mexico. he developed a taste for Margaritas. He drank four of them, one for each man that was replacing him. It took that many men to replace him because they did not have his skill level. But the pay difference more than made up for the difference. Also, there were no employee benefits at these plants or unions.

With him was another laid off employee who had 32 years of service operating a battery truck equipped with an automatic line painter which he used to paint the lines in factory bays as well as the plant roads and parking lots. In the old days, that was his entire job. He admitted it became dull but his last act when he got his layoff notice, he decided he had to have the last word. He used the painter to write his name in the huge parking lot by Bldg. 273 which was not jammed with cars like it had once been.

Even though GE was successfully operating plants in the south without unions and in some cases (*thanks to the Reagan era in which the NLRB was told to back off*) had decertified some, it was still cheaper to move to developing countries.

One of the most advanced lamp plants built right before Jones retired was in Winchester, Virginia. However, it was harvested and moved after GE Capital purchased 51% of Tungsram in Hungary for $150 million. The Welch PR team announced this deal -- along with the purchase of Thorn Lighting in the United Kingdom -- made GE the No. 1 light bulb maker in the world. What they did not announce was that GE was already the No. 1 bulb maker. Now this business dating back to Edison's invention of the incandescent bulb was No. 1 without jobs in the U.S.

It was the selling of advanced technology which got the technical side of the company most upset. For years, the power generation business had offshore business affiliates in Switzerland, Italy, Germany and England in which license fees were paid to use GE technology but GE still kept the rights to its advanced technology. GE's power generation businesses thrived year after year on these fees.

However, under Welch's leadership, these agreements were ignored and the technology sold outright in co-production agreements for huge sums to European and Far East competitors.

I got firsthand experience with one of these sales at an international meeting for power generation people from all over the word held by Alstom, a French energy conglomerate. They had invited 240 customers from 42 different countries to attend their Congres Turbine a Gaz. GE was asked to present technical papers on its powerful MS9000F gas turbine which had been developed for international customers. We had half of the two and one-half day program. I don't think I ever worked so hard in my life. It was a multi-screen projection. They wanted our slides a month early so the French film team they had hired at Cannes could automatically show them without interruption.

We had six talks led off by Don Brandt, our manager of design engineering. Brandt was not only a brilliant design engineer but he loved classical music, too. So we designed his talk around Beethoven's unfinished Ninth Symphony noting tongue-in-cheek that he was finishing the ninth with the most advanced technology to date, the 9F. He called the 9F the Ode to Customer Joy as he outlined the gas turbine superiority in efficiency and reliability. He told them there was not a note out of place in the 9F...that this machine would give the world's best performance. The French film team loved it and added Beethoven music.

Brandt had already endeared himself to the French when he asked me to research information on Carnot, who had been an early French developer of gas turbines. When I went to the Union College Library in Schenectady I could not believe what I found. As a well-endowed college, many of the early pioneers of GE had donated their libraries to it. One of them had given his collection of Carnot's original designs for gas turbines to the library. Looking at the sign out card, I realized it had not been checked out for 40 years. Brandt was stunned when I showed up with the drawings. He immediately told Alstom officials to ask the college for these historical records. *(When I went back to return them, I checked the shelves to find out they also had first editions of three Mark Twain books on the shelves to be checked out. In this case, I informed the State University of Albany about it because unlike Union, they had few books in their rare book inventory.)*

Anyhow, working with one of the engineers to put these presentations together, I think we went through every photo and slide that had ever been taken in the gas turbine business. I also added my usual inserts of GE's history of innovation from the time of Edison. We met the deadline. Just in case the slides did not make it by DHL express, I made an extra set I would carry with me to the meeting. They filled two suitcases.

Despite my marketing communication manager seeing a trip to Cannes as his prerogative as manager, Patterson and Brandt insisted I go because I knew the program. Since engineering was paying for it out of their budget, my manager had little choice. My manager laughed when I told him I was carrying an extra set of slides with me saying that would look unprofessional, reminding me that the French already used the slides I sent.

As it turned out, those slides resulted in me getting a management award from Patterson. When we got to rehearsal, we discovered that all identifications of GE producing the MS9000 had been taken out of the slides. We were told that a co-production agreement was underway -- that Welch was selling the advanced gas turbine technology to them. Patterson was furious and since they could not produce a document at the time the deal had been done, he insisted they change the slides back to their GE identification. The Cannes film team said that was impossible.

Then I mentioned I had a duplicate set. Patterson told the Alstom officials we were going home unless they gave GE recognition for developing the MS900F. With a full day and a half program about to vanish, they agreed.

It was one thing to lose your competitive edge by not investing in product development but to have the advanced technology you already had being sold was too much for many engineers. Many of our best engineers left the company when they saw the business was being harvested.

It was this adage of "Be Number One or Two" or you are "Out" that a product development laboratory chemist named Mark Markovitz got into trouble with Jack and his Gold Collars. When Markovitz read this edict, which he dubbed bumper sticker thinking, he would take issue with it, often writing a letter to the editor of business journals which were touting every word from Welch as the messiah of business.

It was his letter to Fortune magazine that first got him in trouble with headquarters. Markovitz argued that the gas turbine business -- which was Number One in the world now and making 30 cents on the dollar didn't start out that way. It took 10 years, the 1972 Northeast blackout and the increased demand from Saudi Arabia oil barons for GE's new advanced gas turbine lines before they made serious money. Now it was pouring in. Markovitz also pointed out GE would not exist if it had started out with an edict of being Number One or Two or you are out. He also raised the question of was it good for GE in the long run to become a short-term thinking financial company.

I was in Paterson's office with a speech draft to the utility executives when the call came in from the Vice President of Human Resources to get rid of Markovitz. Patterson told me to check him out. He had learned not to trust Human Resources when he found out a tap had been put on his phone. So off to the lab I went where I met one of the nicest guys I had ever interviewed and definitely the smartest. He held numerous patents and his immediate boss had nothing but praise for him. I went back to Patterson and explained the situation which made him even more irritated with the Human Resources VP.

Many technical managers like Patterson resented the fact human resources had been given so much power under Welch to make employment decisions. Too often, they would abuse their power and cite new government rules like EEO as the reason they had to have the final say on who was hired or fired. They did not have the technical background to judge merit and too often their decisions were based on personality, not skill, or as Patterson would say, sucking up to them.

Anyhow, I remembered from my days in employee relations before it was renamed human resources *(resources to me implies employees are just inventory)* that there was a woman named Wilma Soss who used to go to the annual company meeting and do her whistle blowing there to make her point. In doing so, she got national press. She also knew she should go to the meeting every year because employee relations did not want to look like they were retaliating and would wait a year to fire her. Patterson immediately told me to start writing letters for Mark but I told him he needed no help. Under that lab coat was an economist that Reg Jones would be proud of.

It worked and I think Mark rather enjoyed his job security "cover" because he even wrote letters every six months or so instead of once a year. But as Markovitz said, Welch provided him with lots of material to comment on. In coming years, he was to find plenty of bumper sticker slogans to analyze.

When Welch claimed to be the one that put GE on the map globally, this time it was Dr. George Sarney, manager of Gas Turbine

Marketing, who got into trouble. He forgot corporate politics in discussing the history of GE in a business press interview.

Welch's new Vice President of Public Relations at headquarters, Joyce Hergenhan, an MBA, in "yessing" Welch, had given an interview to the national business press noting that GE had not been a global company until Welch became CEO.

When asked about international sales, Dr. Sarney made the mistake of pointing out that gas turbine's early dramatic growth could be attributed to offshore sales starting in the 1970s, particularly after the oil embargo in 1972 and the growth of the oil fields in the Middle East. He added the company had a division called International GE that had its headquarters in NYC in the 50s, 60s and 70s. He even went back to the days when Edison had tried to use a Japanese bamboo shoot as an early filament for his light bulb. That was the end of Dr. Sarney. However, his prowess as a marketing manager was well known outside of GE. He was hired as head of Raytheon.

The 1990s saw the arrival of a whole new slew of Gold Collar management, mostly MBAs, which signaled the final harvesting of the Schenectady plant businesses. Back in the 70s, we had six major businesses in Schenectady. The Large Steam Turbine and Generator business, Gas Turbine, the Small, Medium and Large motor businesses and Power Tube. Each had a Vice President or General Manager responsible for that total business. Now we had one combined turbine business and six vice presidents competing with each other with titles like VP of Power Systems, VP of Customer Service, VP of Sourcing, VP of Materials, VP of Human Resources and VP of Support Operations.

Every time Welch demanded more profits, new titles and re-organizations were announced. The people printing the stationary with the new acronyms were going nuts. In fact there was one that only lasted a day. The VP of Power Systems Support Operation did not like being called the PSSO head.

Each one of these Gold Collars was charged with improving the earnings in their area of so-called expertise which often meant they competed in their drive to find cost savings. As a result, one hand did not know what the other was doing as each one's drive to achieve cost savings (*and boost their Incentive Compensation stock options*) would affect the total operation. Production at times came to a standstill when materials were missing. An order won without regard to operating costs was the most devastating. One VP did not know what the other one was doing to boost his stock options and bonuses.

Making the situation worse was the breaking up of the manufacturing organization into production companies in which each unit had to compete with offshore suppliers.

When the Gold Collars finally left Schenectady in 2000, the plant was down to 1500 employees, the large steam turbine business was gone forever and other power generation businesses downsized or moved out. Ironically, the steam turbine business was so badly harvested, Siemens was no longer interested in buying it. It simply went out of business.

Today Shanghai Electric is the No. 1 manufacturer of steam turbines followed by Siemens and Hangzhau. Hitachi is more like the GE under Reg Jones – being number one in the total range of electrical products serving the consumer, industrial and power generation markets. Since 1985, Hitachi has been leading in U.S. patents. Siemens overtook GE in the late 80s by employing three times more scientists and engineers.

Amazingly, Wall Street security analysts and the business press continued to hail Welch as the business messiah. As long as he made profits, they never questioned how he was doing it. His word was gospel. No one seemed to notice he was tearing down the house and legacy of Edison. But we sure did in Schenectady. Next: How he and his Gold Collar cronies ran the scam.

HOW THE GOLD COLLARS SOLD US OUT

Of all the programs Welch introduced to "Become No. 1 or 2 or Fix, Sell or Close," the most devastating to the Schenectady GE plant was the Six Sigma program.

Six Sigma originated as a set of tools and strategies developed by Motorola in 1986 to improve manufacturing processes and eliminate defects. A Six Sigma process is one in which 99.99966 % of the products manufactured are expected to be free of defects (3.4 parts per million defects or less) It involves a lot of paperwork to define, measure, analyze, improve and control selected projects.

Motorola attributed the success of its launching of the Iridium satellite constellation, one of the most complex projects undertaken by a private company to its Six Sigma program. It involved some 25,000 components and took 11 years to develop and implement at a cost of $5 billion.

So Six Sigma became the rave of many CEOs, particularly Jack Welch, who saw its popularity as a way to achieve another feature of the Six Sigma doctrine which set it apart from previous quality improvement initiatives. It was a way to focus on quantifiable financial returns from projects and its special infrastructure of passionate Black Belts and Green Belts to implement the approach appealed to him.

When Jack introduced it to GE in 1995, it could have been perceived in Schenectady as good news if it was implemented like it was at Motorola. However, since he took over in 1981, profits from the Schenectady plant had never been returned for re-investment in plant and equipment. Unlike Motorola, which invested $5 billion in its Six Sigma project, the Schenectady plant was not only considered a cash cow for GE Capital, but it was not in Welch's three-circle strategy. In fact, by 1995 the plant had lost over 15,000 employees since he became CEO and would see another 7000 go before he retired in 2001.

Six Sigma quickly became known as Sick Sigma when it was found quality had nothing to do with the program. It was a Machiavellian ruse not to fix the business but to sell the technology first and then the business.

Under this program, Welch said he was developing top management by designating the best people in the company as "Black Belts." This meant those following his vision and excelling on Session C performance ratings would hold two-year assignments across the company to qualify them as Black Belts for leadership.

Often those Gold Collars coming into Schenectady on such short assignments knew little about the power generation business. So it became chaos because there was no clear leader aware of the total business as titles were handed out like karate chops. Life turned into a living hell for engineering and manufacturing people who saw each Black Belt compete to wring cost savings and productivity gains out of the business.

As noted, Welch between 1981 and 1985 had boosted stock options from $6 million to $52 million. These incentives increased even more during the late 80 and 90s. It was not just GE Capital Gold Collars who were being outrageously paid and awarded incentives. One of the most incredible examples came when Welch offered Jerry Seinfeld $100 million in stock options in 1998 just to stay another year – and Jerry turned it down wanting to spend more time with his family.

Under Six Sigma, the incentive compensation plan was changed so that 60% of the bonus was based on financials and 40% on Six Sigma results. The lavish stock option grants were focused on employees who were in Black Belt training. Each Black Belt, highly motivated to earn more incentive compensation and stock options, competed to increase "financials" and get Six Sigma results out of the business – either not realizing or not caring how it impacted the other functions.

Trying to coordinate what the Gold Collar in charge of sourcing was ordering from offshore with what was produced in-house became a nightmare for manufacturing. Often the outsourced parts did not fit during final assembly. Worse, if shipped directly to the customer site for assembly and it didn't fit, engineers had to drop what they were doing and spend time at the customer site to fix the problem. Most of the field engineers had already been laid off.

The Gold Collar with the power to reduce inventory to save costs often did it without checking manufacturing. This resulted in missed delivery times when they discovered materials were no longer in stock. Forcing engineers to standardize designs and designate cheaper materials often resulted in the additional work at the customer site. Such added field and assembly expenses didn't show up in the Customer Service or Offshore Sourcing budgets. Therefore, those Gold Collars did not care because they had already been credited with improving earnings. That was somebody else's problem.

Most devastating to trying to save the business was the so-called re-invention of engineering and manufacturing program.

It involved forming manufacturing into production companies in which each would compete with outside sourcing as if they were separate businesses. Essentially manufacturing became a parts supplier business such as the rotor shop, the airfoil shop, the generator shop, stator bar ship, etc. Offshore suppliers also had the advantage of better equipment because investments had been made to improve productivity. So breaking up manufacturing into production companies was a quick way for the Global Gold Collars to boost their incentive compensation and stock options.

Manufacturing and Engineering Managers rebelled against becoming an assembly business. Getting all these parts on time and to fit would greatly increase cycle times and costs to no avail. Again, that was somebody else's problem because there was no single person in charge who could see the big picture.

So, in a last attempt to prove manufacturing and engineering could boost earnings to Welch's demand and not become an assembly business, they suggested a standardization of both the gas turbine and steam turbine lines where possible to save costs and improve delivery time. They did this even though it was obvious the profits would not be reinvested in their businesses because Welch was on a buying spree to expand GE Capital.

The first standardization program was taking 15 designs for air-cooled generators and reducing them to seven. This was highly successful and the generator business thrived and survived on its own for many years.

However, with so many engineers continuing to be laid off, it became an impossible job to quickly standardize other components to prove the business could turn in the hefty profits Welch wanted. Also trying to standardize airfoils was not easy because customers used their turbines in a particular operating mode which justified a particular gradient. Moreover, standardizing the airfoil shop would require an investment in new tooling. Only ideas that saved costs were accepted. A manufacturing supervisor had already gotten into trouble for saying two Black Belts had just received bonuses for taking costs out of the business which could retool two bays of Bldg. 285. Steel was now coming from Japan, machine tools from Europe, copper from South America and furniture from Canada.

Other standardization programs in design never got a chance to prove themselves because gas turbine manufacturing was moved to Greenville, S.C. and steam turbine was barely able to operate with outdated equipment. Even the union had agreed to the re-invention of manufacturing by agreeing to standardize their job classifications which again meant less take home pay. Job classifications of 250 were drastically reduced. The top five rates were for 800 machinists. The midpoint had three rates for the 115 winders. Three rates were available for the 40 wood and plastic fabricators. A total 995 factory workers with 11 rates was a far cry from the days when there were more than 10,000 factory employees back in the 70s with over 500 rates.

By far the quickest way for the Gold Collar in charge of operations to cash in was to tear buildings down as quickly as he could to get them off the tax rolls. So many buildings were being torn down that the plant started looking like a golf course with an inordinate number of sand traps from the air.

That became a living hell for me. In just one year, I had to be moved five times to avoid the wrecking ball. We were in the computer age and it would take weeks for the computer to be re-installed. Fortunately, I had a PC at home and could do much of my writing there.

But the insanity undoubtedly was worse for the engineers. A new DC lawyer had been added to the plant to make sure there was no price fixing going on (*Welch was not about to have another charge of price fixing*). The lawyer decided the most effective way of making sure there were no outside purchasing prices that looked suspicious was by getting rid of all the engineering records which filled several record rooms.

These records of every turbine sold went back to the 1900s and were very useful in a growing upgrade business for utilities during planned maintenance shutdowns. Engineers pleaded to save these drawings and records for future business. Obviously, the lawyer did not have the time or knowledge to go through such a massive amount of information. Besides, that was long range thinking and stock options were awarded each year.

So when he ordered the records be destroyed to assure there were no records on pricing, I was not the only one taking things home. Engineers were seen for weeks loading their cars with file folders -- while hoping the Gold Collars would leave town once they got their pockets filled.

The market for new unit orders was sluggish but reserve margins had been eroded enough to force utility managers to go to their commissions to get funds approved for upgrading and adding capacity. Fortunately, history had taught the engineers, who thought long-range, this day was coming and they had saved these records. Those who bought service shops when half of them were sold by Welch's Gold Collars were even happier. Their quick purchases paid off for them big time.

Not being able to list nearby GE service shops as another competitive edge was also hurting the business. The days of charging a premium price because of advanced technology and unmatched service were over. Yet, Welch's drive for profits at any cost while maintaining a No. 1 or 2 position was intense. So intense that when the turbine business started losing market share to ABB and Siemens, a guy I knew who followed market share, Ross Bradshaw, reported it. When he refused to change the numbers for the Gold Collar Marketing head, he was fired.

The first Black Belt Gold Collar designated as top management to move in to get costs out of the business was Russ Noll. He was sent in to replace George Cox as Vice President of the Steam Turbine and Generator business. Cox had resisted the heavy layoffs, citing the damage it could do to meet delivery schedules while pointing out the business was making a profit -- albeit not as much as Welch wanted. This contradicted the usual "climate setting" announcement from Welch's PR staff when a new Black Belt manager came into Schenectady. Every two years the mission would be the same: "turn the business around and make it more competitive."

Noll proved to be very adept at a "divide and conquer" strategy by keeping the union informed of what was happening in the Session C meetings to solicit their support. He befriended the union leaders to let them know the office people were going to have to learn to be more competitive. He often lunched with the union representatives and dressed like them.

Even though outsourcing was causing many blue collar workers to lose their jobs, Noll could easily point out that the vast majority of his reductions were in reducing the white collar ranks. He succeeded in laying off 30% of the office workers over the next two years. Noll would point out that those rising tides that floated all boats days were over. He would remind the union leaders that white collar office people had always enjoyed the benefits of increases union folks achieved without having to pay dues.

With the help of Welch's Human Resources VP Gold Collar, Noll began red circling white collar jobs held by those who were doing work that could be done by new entry-level employees at a third or half the price. This was a reminder to the factory employees of the days in the Make Schenectady Competitive campaign when they lost piecework pay and saw many jobs reevaluated to lower wages.

A new atmosphere of fear took over with the Red-Circle program. It got so bad that employees, who had been red-circled, were ostracized by other employees fearful of a close association with someone who had been deemed as non-competitive. It was particularly tough on long-service employees who had been very loyal to the company and felt their experience counted for something.

Rebuilding the GE House Jack Blew Down

Some were able to find other jobs outside of GE. But most were too old and had to settle for less pay if they were not eligible for an early retirement.

I remember a couple in their late 30s who had just bought a house. He was a payroll clerk who was very popular in the plant. For years his wife had done programming work at less pay than those with degrees and had outstanding job appraisals. She had been going to college at night on tuition refund to bring her salary up to par. She had just achieved the raise when her husband was red circled and despondent. They decided life would be better for them if they cashed in their Savings and Security funds and bought a camp in the Adirondacks. But his depression was too much for him even though the camping business was growing. He left town without a word to anyone and today no one knows where he is. She ran the camp for years and is now a leading programmer for a bank in Albany.

There were also many scientists at the Research and Development Center who could not believe what they did was no longer important. When Welch stopped basic research investments to save costs, many of these scientists got pinks slips. I remember one of them who had contributed heavily to semi-conductor research being stunned when he got his lack of work notice. He was so well regarded in the field, he proudly told the Black Belt manager he had written the definition for semi-conductors for the World Encyclopedia. The company had not been paid for it. He was able to take an early retirement.

Some found jobs at universities and companies still interested in R&D. But there were a couple of scientists with such pride in their work *(and obviously no outside life to take up their interest)* that they actually committed suicide, leaving notes.

It was painfully clear that if you were not a Black Belt Gold Collar and followed Welch's financial services, you were not going to get bonuses for just doing a good job. Past performance meant nothing. Welch made this clear when he found out there was a highly paid Level 19 individual contributor at the Research Lab who had worked on the Manhattan Project and gotten national awards for it. Even though he was now doing research on the effects of radon, he was quickly retired and used as an example of how you cannot rest on your laurels – that what you do today is what you are judged on.

So walking Welch's talk and becoming a Black Belt was the way to achieve financial success. If you didn't meet that criteria, then becoming a Green Belt (*thousands of employees who were more than willing to go along also went to Crotonville for training*) was the next best profitable step.

In the first full year of Six Sigma, some 30,000 employees companywide underwent training of Jack's vision at a cost of $200 million which Welch claimed paid off in later years. These sessions included a game called "Constructive Conflict" in which the Black and Green belts would assume manager roles and compete for the best cost-cutting and money-making idea. It was during one of these sessions that Welch learned how valuable the investment of the $80 million in the new 7F and 9T gas turbine started during the Jones era was. One Green Belt noted that Patterson said high efficiency and reliability tests had shown the technology designs were so advanced that it would take competitors at least 10 years to catch up -- thus assuring long-term profits for that business. It was then that Welch saw quicker and higher short term earnings by selling this technology through co-production agreements with offshore competitors. Green belts were not just useful as cost cutters, they were also sources of information.

Such Six Sigma sessions for Black and Green Belts not only allowed every stone to be turned over for costs savings, but it also created a class of "yes" men and women who spent their time being political. They were more than happy there were those still left who took pride in doing a good job. They wanted them in their organization until the harvesting was completed.

Thus, doing a good job and not having the Black or Green Belt label also meant you would never be promoted by them. Certainly the best high-profile example of this was Dave Letterman who followed Johnny Carson with the late show. When Carson retired, it was assumed Letterman would take his position because his ratings were so high. But the GE Gold Collar running NBC saw it differently. He hired Jay Leno to replace Carson figuring he had two winners bringing in big ad fees. Dave jumped ship. All of us cheered when he did knowing someone had not been taken advantage of for doing a good job.

But it was not just taking advantage of people and saving on payroll costs that improved earnings. To save operating costs, nothing seemed to get by Noll. One day he noticed all the water coolers in the offices. This terse Oct. 11, 1988 letter was sent to managers: "Please assure removal of bottled water from each of your components. It's an unnecessary expense since there is plenty of city water available for drinking."

The guards were the first to object when the water bottles were removed from their stations. They would have to wait on a replacement from the main security office when they needed to go to a nearby building for a drink. The same was true for many office and factory workers who did not work near a fountain with city water. Many also remembered the days when there were signs in the johns in Italian and Polish that it was not safe to drink the water there.

However, it can be said Noll was consistent in his cost-cutting. He did not redo his office with new furniture as most Gold Collars did when they arrived at the plant. He was also a patriot in selecting an American car for one of his perks -- unlike other Gold Collars who chose expensive foreign sports cars. However, it was his zeal in cost savings that got him into trouble with Welch.

His first mistake was to save $20,000 a year by shutting off the huge General Electric sign at the plant entrance stretching across the top of Building 37. It used several thousand light bulbs which also had to be maintained. When a Wall Street Journal reporter asked if the lights were going out in Schenectady under Welch, they were quickly turned back on. So Noll relented and turned them back on. But he still saved some $5000 by not replacing some of the bulbs with green and red ones for the Christmas holidays.

It was another bad public relations move that got Noll into the most trouble. The Number One challenge Noll had listed in his "Bold Strokes" vision for the Schenectady plant was to "Become the Low Cost Supplier." A top customer noticed the huge Bold Strokes banner at the front of the plant and had questioned Welch on whether the company was still committed to quality. But instead of redoing the banner which would have been costly, Noll simply had plant maintenance go up and arrow in the words "quality" before "low-cost." This resulted in a new picture run by the press with comments of quality being an afterthought.

Welch's Human Resources Gold Collar moved in to remove Noll. His replacement was to make employees wish Noll was back. While all management came in with the announcement they were here to "turn the business around" (*which had all of us dizzy because it was happening so frequently*) Noll did appear to want to make the business more competitive and profitable. He set the example of not wasting money on plush surroundings for himself.

So when he was followed by the biggest Gold Collar of them all, David Genever-Watling, it was a shock. The first thing he did was build himself a $7 million dollar office on the top floor of the old General Engineering Lab building which had the huge lighted General Electric Monogram on top. A private elevator was also put in with a Pinkerton guard at the entrance. Naturally plant security people were insulted because they had a history of providing good security. It was only natural they would wonder how many years of bottled water could be paid for by not hiring a Pinkerton agent.

Genever-Watling was quickly dubbed two-names or sometimes Dash when people noticed his British coat of arms emblem displayed on his jackets. Speculation had it that Welch decided he needed a guy that exuded quality at the head of power generation to impress customers. Engineers would point out utility heads were not MBAs -- but mostly electrical engineers who came up through the business and, therefore, would be very unimpressed.

A grand entrance to Bldg. 37 included a state-of-the-art video center and auditorium to enable two-way conferences within the company as well as with customers. Presumably this would save travel costs and save time. A café was added. But only the Belts seemed to use it. Others felt anything they said could be used against them, particularly if it involved customer projects which might reveal a way for the Belts to suggest ideas to save costs which would further hurt the business.

Some 254 parts and orders catalog people had to be shoved aside to make room for the top and bottom floor of Bldg. 37 now taken up with Dash's office and the grand entrance. They were jammed into cubicles so tight that Chuck Klages said you had to be a Houdini to get into them. Though he did not have a bachelor's degree, he was admired by one and all for his knowledge of turbines parts from flange to flange. During the move, there was enough money left over from the video center to purchase a computer system to automatically order parts without having to look up the item in the catalogs which were the size of the Manhattan telephone book.

Walmart, which had used such a computer system, was hired to send in experts to train everyone how to use them. I don't think I ever saw anyone as disgusted as Klages was with the whole process. He noted that the Walmart people soon found out ordering turbine parts was a lot different than stocking panty hose. They left claiming

GE had bought a cheaper model than the system they had. Klages said the software program for their system must have been designed for a mop and wringer company. Fortunately, the parts people had done what the engineers had done. When they were ordered to get rid of their parts catalog, they took them home. And they stayed there because as Klages said, there wasn't room in their Houdini cubicles. Still, he used his cubicle computer for correspondence.

Even though Klages decided there was no conscience among the greedy Gold Collars to tweak, he used to secretly post thoughts for them to discuss on the way to the bank. His favorite was to ask: "Does it bother you our offshore competitors can soon raise their prices knowing that we no longer have our own steel foundries, copper wire mill, tool and die operations or other in-house capabilities? Are you making enough harvesting the business to take care of your kids in the future when jobs will be scarce?"

Because there had been so much reduction in employment and buildings torn down, finding a parking spot in the plant had certainly ceased to be a problem. Still, when Genever-Watling had plant maintenance line out a spot for his chosen perk car, a Ferrari, in front of every major building, it caused quite a stir. Moreover, he had the two spots on either side blocked out just in case someone got too close to his car and dented it. The guards renamed him Double-Perk. To say he was disliked would certainly be an understatement.

My 60-hour weeks included getting to know the cleaning staff at night. One of these was a tough Irish woman who had seen her job go from cleaning two floors to three buildings. One night while I was in the john she showed up with Genever-Watling's monogrammed coffee cup. She showed me the note his secretary had left complaining that it had not been cleaned properly. I heard the toilet flush and she came out with it, asking me if it looked clean now. I must admit I laughed and did not advise her to rinse it thoroughly.

This perception of Genever-Watling being a dandy was certainly born out in the first meeting he had to address utility executives across the country. On a personal level, I had already questioned his know-how when he complained that the slides we had of turbine and generators were dull looking. He never questioned the ratings but asked for some red, blue and green ones. So all my work with engineers to make sure customers saw photos of their best units at the rating they liked was for naught.

But his arrogance hit a high when he inserted in his speech what he thought was a profound phrase – "a megawatt doesn't care where it comes from." Even one of the other Black Belt managers suggested customers might not understand his "deeper" meaning. But he went with it. The customers looked at each other and decided that maybe Noll's vision of being the low-cost supplier was still in effect...that it didn't matter if the equipment was of high quality as long as it put out an "uncaring" megawatt.

Making it even worse was we had tried to count up maintenance costs as if they were investments (*including his $7 million office*) in hopes of showing customers we were going to maintain our competitive edge. Genever-Watling stated $7 million instead of $17 million. He looked confused when a customer at the end of his talk questioned such a low sum in plant and equipment. Genever-Watling didn't answer him but repeated his pleasure at meeting with them.

It was particularly discouraging to employees who saw more layoffs along with the redcircle program continue under Genever-Watling's reign. In the beginning there had been fairly good retirement incentives to get people to retire if they were eligible. But now these were starting to be reduced. While many hi-potential, young engineers were able to leave early on when they saw there was no future, now older engineers with solid reputations were sending their resumes out. The "turn the business around and be more competitive" story was no longer being bought by the vast majority. The onset of the Gold Collars, who based all decisions on quick payoffs of stock options and cash bonuses for them, had taken its toll on credibility from headquarters. The "Fix" part of Welch's Fix, Sell or Close edict was now replaced with "Harvest."

When word of what Jack Welch was spending at headquarters and the company's management training center went around, the need to improve earnings through productivity gains mantra sounded even more false. Welch spent $75 million for a fitness center, guesthouse and conference center at headquarters and upgraded the management training center which he described as "pocket change." He said the "improvements were as much about getting people together as it was about health"...that he wanted to "create a place where people could stay, work and interact."

Unsaid, was the fact that not only did Jack have 18,000 people going to the training center in Crotonville to pass his vision test, but now he had a loyal staff that he could also control with A-team perks.

So while Genever-Watling's $7 million office might look like "pocket change" to Welch, it did not to engineering and manufacturing people begging for new tools and equipment.

Genever-Watling's actions made him even more unpopular. To assure himself he had total control, he demanded that any transaction with customers had to have his approval. Customer orders often involved a penalty if not delivered on time. A part that needed to be shipped to a customer in the Far East overnight had to have his approval immediately to get there in time to avoid a $50,000 penalty. The manager sent a fax to Genever-Watling for his approval. In those days the fax copy came out on grayish paper. Genever-Watling refused to sign anything that was not on 100 weight white bond paper. By the time the GE supply foreman found the correct paper for Genever-Watling to sign, it was too late for the DHL shipment. This costly $50,000 "hissy-fit" did nothing to improve the morale of the field assembly engineers on site.

While Noll and Genever-Watling had done their jobs of improving profits at any cost from tearing down buildings to get them off the tax rolls to layoffs to outsourcing, it was the arrival of Gold Collar Bob Nardelli that signaled the end of the business in Schenectady. He kept a low profile but was well known as Welch's pet Black Belt Gold Collar.

Gas Turbine was moved to Greenville, South Carolina and to Nuovo Pignone in Italy. Nardelli was also useful in brokering co-production deals to get quick money on advanced gas turbine technology.

Steam turbine was now being moved outside of Welch's three circles, which many said meant he had given up trying to sell it to Siemens. When that rumor was first run in the Wall Street Journal, an engineer had immediately faxed his friends about it. Human Resources obviously did not read the Journal as early as he did. He immediately called for his dismissal before realizing he had not started the rumor.

Many engineers and manufacturing people hoped Siemens would buy it because their power generation business was being supported by investments in equipment as well as research and product development. But Siemens knew all they had to do was wait. The word was out that steam turbine was considered a commodity by Welch. Why buy a harvested business when you can just watch it go out of business? Instead Siemens bought Westinghouse and ended up with most of the U.S turbine business.

GE used to be the No. 1 leader in power generation and delivery and, as noted, today it is Siemens, Shanghai and Mitsubishi.

Schenectady reached a new low of 1500 main plant employees under Nardelli's reign.

Noll had improved earnings by getting costs out. Nardelli finished the job started by Genever-Watling to make money going out of business by selling the technology and what was left of the product businesses in the plant.

Nardelli also liked the lavish style of Genever-Watling although he did not flaunt it. However, one has to note that when he left GE to become CEO of Home Depot, photos of his 12-car garage and opulent home financed by Home Depot were a big hit in plant e-mails.

In retrospect, there was some sympathy for Noll. Many felt he had believed in the business and could produce the profits Welch wanted from the business. However, when Noll was replaced and retired by Genever-Watling, Noll told the local press that he had received $2 million in incentive compensation that year so some of that sympathy was subdued. Noll was also probably dropped as a Black Belt because Welch found out he did not share his global vision.

A union leader told me that he believed Noll when he said he rebelled against doing a coproduction deal with China because there would come a time when China, after luring billions of dollars' worth of U.S. manufacturing business into their country, would simply lower the Iron Curtain.

The mortal blow for many other GE businesses across the country occurred when Six Sigma was introduced in their Product Development Laboratories. These labs, serving the business strategy of having the best and latest product innovations, were the seed corn for future growth through customer satisfaction. The Six Sigma program to "define, measure, analyze, improve and control" each project was particularly counterproductive to technologists.

Here's how Mark Markovitz, known for his patents in superior insulation materials, described the effects of the Six Sigma program: "Six Sigma placed a thick, bureaucratic shell over the scientific process. It wasted many millions of dollars at GE. It created cynicism because engineers' compensation was based on completing Six Sigma projects and the engineers knew the projects wasted their time. Because a number of Six Sigma projects were required each year, the projects usually started as engineering projects that were completed efficiently using the scientific method but were then formatted for Six Sigma to allow the engineers to receive credit and kudos for using Six Sigma. Many hours of engineering time were wasted for formatting to Six Sigma graphs and tables. Additionally, hours were spent to obtain additional irrelevant data to widen the data base for Six Sigma presentations – the additional data did not add any new knowledge to the engineering project that were first completed quickly using the scientific method.

"Fishbone diagrams used in Six Sigma contain all possible possibilities, many far-fetched. This prevents rapid focusing on the most probable causes of a problem. A few examples show the waste and fraud that resulted from Welch's mania for promoting Six Sigma in GE. A chemical engineer at a GE insulation conference gave a talk on how Six Sigma made possible a more precise measurement of viscosity of insulating varnishes at a motor plant. He described how he spent more than a month collecting data comparing viscosity readings using the "old" and the "new" methods. These methods can be used on the factory floor. These readings were compared to values obtained with a Brookfield viscometer in a lab. The Brookfield viscometer is not a factory instrument. He showed charts and tables showing how "new" factory method viscosities were more closely aligned to the values obtained with the Brookfield viscometer. Great Six Sigma outcome! However, to those who were knowledgeable of viscosity test methods, it is well known that the "new" method would be more precise than the "old" method. The chemical engineer wasted many weeks collecting data to show something that was already known. In addition, he wasted the time of the engineers listening to his Six Sigma "success" story. It was not the engineer's fault because he had to present successful Six Sigma projects to management.

"There are two examples where I was involved. Generator cores have steel space blocks between sections of the laminations for cooling. The space blocks are dip-coated in a solvent based alkyd enamel. Because of the rapid drying of the enamel, it was necessary to further dilute the enamel with solvent, which increased the VOCs (volatile organic compounds) emissions. Production would soon be stopped in this manufacturing area because the VOC emissions limit was being reached. I recommended a change in the solvent composition of the alkyd varnish as supplied by the manufacturer to a lower evaporation rate solvent. It took less than 2 hours to choose a new solvent based on evaporation rate and alkyd resin compatibility. The problem was solved. No additional solvent had to be added to the varnish made with the new solvent. Production could continue because the VOC emissions were under control. Not only did it reduce VOCs but it also cut costs because drums of diluting solvent were no longer necessary. The problem was quickly solved in two hours. However, this work had to be presented as a Six Sigma project by another engineer who spent days converting the work to the Six Sigma format. Many tables of solvent properties were added to fit the format.

"There was a major stator bar manufacturing problem concerning the high voltage insulation. Stator bars see very high voltages and use mica-glass tapes containing a heat resistant epoxy resin for insulating the conductors. The stator bars are manufactured in an autoclave where the stator bars are first vacuum-heat treated to remove air and volatiles followed by curing under pressure. After the bars are made, they are tapped to detect any porosity in the insulation. There were a large number of bars that failed the tap test and had to be stripped and reinsulated. This is an expensive problem because of the rework necessary.

"In addition to the problem just described, there were two additional problems. GE Global R&D people were called in to consult on the cause of the problem. Many meetings were held and many fishbone diagrams were generated. Many tests were being done and many changes were tried, all unsuccessful. I was retired at that time but was hired as a contract employee to work on a new product. When I heard of the problem, I knew that overheating during the vacuum cycle would cause the problem. I knew that manufacturing QC measures bar dimensions and differences in bar dimensions would result from overheating. From the bar dimensions, I showed that there were differences in bar dimension for the "good" and "bad" stator bars; there were two populations. I also showed that the second and third problems were also caused by the too-hot-vacuum cycle. I also explained how a change in equipment that took place before the problems began was the reason for the problems suddenly occurring. When I communicated this to the Black and Green Belts working on this problem, it was rejected because it did not meet Six Sigma requirements for data collection. Also, the three problems were being treated as three separate Six Sigma projects, with each own set of fishbone diagrams and data collection. For two and one-half years after I presented the too-hot-vacuum theory as causing the problems, no changes were made to the vacuum temperature. Many millions of dollars were wasted in rework of stator bars. When everything else failed, finally the vacuum temperature was reduced and the three problems were solved. Even when the Black and Green Belts were told the probable cause of the three problems was caused by overheating during the vacuum cycle, which in turn affected the flow properties of the resin during the pressure cycle, which then caused the three problems, the Belts were not able to connect the dots to take any more action after two and one-half years.

"Because Welch promoted Six Sigma, management promoted it and reported its 'successes.' Except for the obsequious managers, most managers and engineers knew it as an expensive waste of money and time. I wrote to the Ombudsman concerning the fraud and waste of money. Instead of responding to my concerns about the program, management avoided having any contact with me. I no longer received an annual review of my work because it would have allowed me to raise the Six Sigma fraud issues with management. I retired soon after that. Although I worked at GE for 38 years, there was no exit interview. An exit interview would have raised the issues I wrote about to the Ombudsman and management would have to confront the lies and fraud of Six Sigma."

Markovitz was just one of the engineers lucky enough to be able to take an early retirement. Today he is still working in his field for companies that work efficiently he said because they are not burdened by Six Sigma.

Some of the technologists at the Research Lab like Bill Kornrumpf, who was part of the team ready to revolutionize the appliance business with digital technology, just did the extra work to incorporate their ongoing projects into Six Sigma. Here's how Kornrumpf explained it: "I did do my required two Six Sigma projects and proved that I could save the R&D Center money. But like all the others, I just took one of my projects and formatted the results in the Six Sigma to get the thing approved. All the program and lab managers had to do one or two. They were really put up jobs. My immediate manager just had one of the secretaries tally up all the unapplied hours for his group and then he did an analysis that showed he could reduce the unapplied hours if he could remove the fixed unapplied time (*vacations, holidays and sick time*) and put them in the overall R&DC budget since he couldn't control that spending anyway. It turned out that most of the managers came to the same conclusion. I never heard of a Six Sigma project that actually was useful."

For those veterans, who had developed technical reputations, finding other jobs was not difficult as consultants or in forming their own business. Some took advantage of the federally financed retraining program. I remember one who ended up in the retail business. He told me his only joy in life now was his family. He added that he is just a nine-to-fiver now but the good part is he is able to spend more time with his family. In the good old GE days, he never minded working extra hours to get a job done. What made him even more disgusted talking about GE was that no matter what they did to meet profit demands, the business was doomed. It became apparent Welch was harvesting the business not only for the benefit of himself and his gold collar managers but to finance GE Capital as the new GE.

It was during Six Sigma that one saw a mass exodus of technical people, some of which had enough service for a vested pension if not early retirement while others laid off. I know a group who meet on Friday's at a local bar. It is little consolation to them that some analysts are starting to see what they saw back in the 90s -- that once GE lost its technical talent in this mad harvesting by Welch and his cronies, it would be hard to go back to product development because it takes years to evolve and develop technology talent.

Time is still telling how disastrous Welch's proclaimed financial vision was. In 1984, Employers Reinsurance was purchased for $ 1.1 billion and the losses incurred by business continued to plague the company until Welch's successor had to sell it for $250 million.

One of the most egregious GE Capital purchases was the Kidder Peabody Investment Banking for $600 million. First, it was found that their hi-potential star, Martin Siegel, was found guilty of providing tips to Ivan Boesky. He paid a $9 million fine and GE paid $26 million in fines. In 1987 things got so bad Kidder lost $72 million and 1000 people were laid off. Not long after that happened, a key trader, Joseph Jett, made a series of fictitious trades in government bonds totaling $350 million which could not be identified in 1994. In the previous year, Jett's phantom trades accounted for nearly a quarter of the profits made by Kidder's fixed income group.

Jett had received a $9 million cash bonus in addition to stock options. Jett was fired. After that fiasco, Welch's drive to bring in even more earnings from the product businesses became even more intense.

When Gary Wendt became head of GE Capital, Welch, declaring every business must take on globalization, said in his Straight from the Gut book that "no one practiced it more effectively than GE Capital." Just in one year, 1994, Wendt picked up more than $7 billion in offshore assets and in 1995, $18 billion offshore. GE Capital was on a global roll, acquiring consumer loans companies, private label credit operations and leasing operations for construction equipment, truck trailers, railcars and airplanes. Ironically, GE Capital businesses did not have to be No. 1 or 2 in their field or high technology for that matter. Among the purchases were companies which were also harvested and retailers like Montgomery Ward which went bankrupt.

The best GE Capital customers in the U.S. were local, state and federal governments in which politicians found it easier for their careers to rent equipment to maintain and build roads rather than buy directly from the equipment maker. A huge single expense in one year would not be good, particularly in an election year.

While GE Capital did make a huge amount of money for many years, it can be said most of the acquisitions to enable it to do so were financed by harvesting the product departments and not putting money into research, product development and plant and equipment.

It became mental torture for many when they read the national press and saw that Welch was receiving wide acclaim as the most admired CEO in the world by security analysts. His Human Resources, Public Relations and legal team had done their yes-job well.

In fact, Welch's ego was so big he even tried to change the name of the company. Even though Welch had gotten rid of most of the appliance businesses which he felt made GE look old and stodgy (*he had a high school survey to back him up*), he felt GE needed a modern name. Fortunately the firm was reputable and did a thorough survey of all businesses, including GE Capital to determine the need. They produced impact statements from the technology businesses that were left as well as statements from major GE customers. They advised against renaming the company but did suggest the company image would be modernized if the "General" was taken out of General Electric and the company just be known as GE. The Victorian style GE logo was re-invented by taking the Edison Victorian era curls out of it and leaving a bold stroke look.

Welch went along with it. The San Francisco firm got its $3 million for taking out 13 letters and a couple of curls out of what we called the GE Monogram meatball. But many more millions were spent on changing everything from stationary to the turbine nameplates to have it read, GE, and not General Electric. One engineer opined that the product side of the business might as well be renamed the Cash Cow.

No one escaped contributing to the bottom line. If a host city or state did not give the company a rate of assessment tax break, then GE would have to move somewhere else. As stated, Mike's hot dog diner in Schenectady had a higher rate of assessment than GE.

Even though Welch and his Gold Collar cronies had total control, one still has to ask, "How could they get the thousands of employees to continue producing at their "A Fair Day's Work for a Fair Day's Pay?" while the Gold Collars were paying themselves outrageous salaries and bonuses while harvesting the business.

The Jones era of loyalty producing productivity gains was gone. Welch felt that fear was a greater motivator for doing a good job.

There are always those who have an engrained desire to do a good job even if they are not rewarded for it – you might call it a sense of self-satisfaction. There were also those who worked hard hoping to get to retirement age before being laid off.

But there was another factor that Welch used to keep the troops going until he had gotten every cent out of the business he could. This was a series of slogans and so called motivational programs full of platitudes that came at such a pace over a 20 year period -- that if collected in one pile -- could fire a waste burning plant for years. For those of us left, we kept our sanity by developing a taste for sick humor. The programs and slogans he used to cover up the harvesting of the business were mental torture for those trying to save the business. Only sick humor got us through the insanity of it all in the 90s. That's next.

SICK HUMOR- DEALING WITH THE INSANITY

Welch's three circle strategy of "Fix, Sell or Close" was as baffling to customers as it was to those businesses in core or outside the circle which were making money, just not as much as Welch wanted. Customers heard about his vision, and were uneasy about doing business with suppliers who might not be around for servicing. Competitors, especially foreign ones, knew all they had to do was wait it out. As time went on, the word "Fix" in the "Fix, Sell or Close" mandate became known as "Harvest."

The harvesting was conducted under Welch's favorite programs called – Work Out and Boundaryless and, as noted before, Six Sigma. With each program came kits full of posters and slogans.

Welch proudly stated in book, "Straight from the Gut," he came up with the idea for three circles when he and his wife were having dinner in New Canaan, CT. While explaining his vision for GE to her, he had used his cocktail napkin to draw the circles headlined "Be No. 1 or 2" or "Fix, Sell or Close."

It was joked he must have run out of cocktail napkins when he added different ways he could boost earnings under the guise of improving quality in Six Sigma program.

"Work Out" was a program he said he came up with while returning in his helicopter after working out at the management training center fitness room in Crotonville. "Boundaryless" came to him while he was lying on the beach in Barbados watching Santa Claus emerge from a submarine. He called it idea sharing.

But it was no laughing matter when the results of these programs went into effect.

"Workout" in the plants was patterned after the sessions in Crotonville in which over 18,000 high-potentials stayed that way if they followed Welch's vision. Thirty to 40 key employees at each plant would be invited to share their views on the business. Those chosen were to represent every facet of the business, including factory workers, financial, marketing, engineering, manufacturing, and security and maintenance people. The meetings could last two or three days *(some of them were held at hotel meeting rooms instead of in the plants)*.

Facilitators, usually Green Belts but sometimes Black Belts, would call the leader of the business they were discussing to give a "yes" or "no" answer to their ideas either right there or at an agreed date. The facilitator would report on "progress" to Welch.

In the case of the Schenectady plant, it was often Jan Smith of Human Resources who ended up heading that organization as a result of her amazing reports. These meetings -- which Welch had quoted one employee as saying "for 25 years, you've paid for my hands when you could have had my brain as well, for nothing,"-- not only resulted in lost time on the job but also became a nightmare for manufacturing and engineering trying to deal with some of these ideas.

The most costly and time-consuming ideas were from financial and marketing people who knew the cost of materials but not the particular traits or applications of such materials. For example, plastic could not be substituted for steel in high temperature applications.

Yet, an engineer or manufacturing person would have to spend hours explaining why the substitution would not work because it could not be machined to the tolerances needed or it would melt in actual customer operation. Questions on why there were so many machine setups needed before final assembly were quickly answered by engineers who suggested this could be done with investments in new electronic and robotic equipment. These comments never made it in the reports because they were not judged as cost saving ideas.

The best way the technical side of the business could keep themselves sane was to go on the offense. Knowing how ridiculous the ejection of the Great Bear water bottle supplier had been, they often suggested that all personal coffee percolators be eliminated and the vending machines that were in the plant should only offer one type of coffee...that it could even be sourced from a third-world country with the lowest bean picking cost.

The suggestion that money could be saved by not sending engineers to technical society meetings was another slap in the face for technologists. However, when the American Society of Mechanical Engineers named gas turbine engineer Nancy Fitzroy as their first woman leader, Human Resources backed off knowing this could result in bad EEOC press.

A goal of getting $250,000 each month out of the business was set by the Black Belt Gold Collar managers. The VP of Human Resources, Dennis Donovan, set the record the first year with $3 million in savings by vending out the cafeteria, mailroom and cleaning services which reduced headcount by 280. He boasted it only took 170 people working for vendors at half the pay to do the job. What he didn't say was that they only cleaned once a week instead of each day.

His suggestion to get rid of the plant library only saved the salary of three employees. But he was allowed an extra cost savings to improve his bonus because closing the library also emptied the bottom floor of Building 2 to make room for employees from a nearby building now slated for the wrecking ball there to get taxes down.

When the library was closed, I found myself and several engineers trying to save research as well as historical records heading for the dump. At that time, we still had a Hall of Electrical History run by the Elfun Society *(an old club for high ranking employees in the paternalistic years)*.

Because Welch had made it clear any business that "was preoccupied with its past" was out, management, particularly from Human Resources, wanted to be perceived as not glorifying the past. Thus, the two original Edison buildings in the Schenectady plant were torn down. Human Resources was well aware of how Welch had made his point about not revering the past when he became CEO. He fired Dee Logan, who had been cited by Reg Jones for doing such a tremendous job commemorating the company's 100th anniversary in 1978. Logan dedicated many issues of the company magazine, the Monogram, to the remarkable growth of the company. While the Monogram no longer exists, one has to admire Dee Logan for an issue before he left which was full of Welch slogans that spelled out the letters in Jack Welch across the center spread. It used each letter to point out his vision. The letter C in Welch was the lead-in for this statement: "Closing a business...." In the old days, the letter C in slogans was usually for customers or competitive edge.

The dumping of the plant library was done on Friday after 5 p.m., the day and time one can always expect bad news to be delivered because employees might forget it over the weekend. However, there were those of us working late trying to save the business who saw the library being emptied and unceremoniously being dumped next to an entrance for pickup by an outside vendor.

Another Work Out cost saving idea had been to lay off plant trucking employees and sell the trucks.

The problem with those of us, who were now literally preoccupied with saving the past, was the volume of materials that had been thrown out. While much of it could be hauled to the Hall of History now located in Bldg. 28 *(not yet scheduled to torn down)*, there was a lot of material not considered historical but good for business. Before the age of computers being able to quickly translate documents, library clerks were hired who could translate Japanese and most European languages immediately into English and vice versa. These records of international business could be important to future upgrade and service business. Because there were only about 10 of us sorting through everything, it was impossible to get it all carried out by car.

Nearby was a covered, but accessible garage for several Gold Collars in that area of the plant whom I knew were in Crotonville for a week of Welch's what we called his "walk my talk" sessions. We stored the rest of what we felt the business would need in there and proceeded to haul it out during the week while they were gone.

But this VP of Human Resources was not through with his $250,000 ideas and the Incentive Compensation that came with it. Next came the layoff of all low-level secretaries which he claimed saved $3 million a year. It was true word processing computers could do a lot of the work. But it was also true most of the investments in computers went to high-level managers, some of which had secretaries to operate them. We laughed about one of the Gold Collars who always left his door open with a view of his computer displaying all the meetings he was attending. No one ever saw him operate it.

One of the most confusing Welch programs was "Boundaryless." Even Human Resources Black Belts couldn't figure out how to put a positive spin on the term. So they just ignored us when we suggested there were no boundaries in layoffs, saving costs, selling technology, moving offshore, forcing tax assessor to reduce rates, etc.

However, it was the layoff of regular secretaries that gave me room for my files.

When the marketing communications staff of gas and steam turbine was combined, it resulting in two employees laid off while I was also to do their work under a program called "Job Enrichment." I was now in Bldg. 273, which once had been called the Grand Canyon of Industry. When it was opened in 1950, it was the largest, high-bay manufacturing building in the world but now its 50-ton cranes were idle. Back in the 60s and 70s we used to brag we produced large steam turbines which could provide electricity to millions of people with the precision of a fine Swiss watch – so precisely built that it could harness steam traveling over 1000 mph with centrifugal forces of two tons on each airfoil.

I was thrilled to be in this building. However, because there were so many other employees being moved in there to avoid the wrecking ball in buildings being torn down to get them off the tax rolls, there was little room. I found myself in a former maintenance closet next to the balcony where customers used to view the progress of the huge turbine generators being assembled below for them.

Because Welch (*often called Neutron Jack*) had nuked the Advertising and Sales Promotion Department, I had moved one of their graphics people, who was now working for an outside vendor without benefits, into my office to handle the orders for slides and photos. This relieved some of my work load by eliminating myself as the middle man.

No one seemed to care Dave Wheeler was not an employee. I was able to get him a guest pass to come in. But we were practically on top of each other in that small space. It was then I decided to take the "Boundaryless" program more seriously. Because there were just two secretaries left on our floor under the Job Enrichment program, I moved a lot of my stuff into the ladies room which was huge. There were 10 john stalls in there along with an outside room which had a couch, bed and chairs in it -- an accommodation from the paternalistic past when it was felt women had to rest.

I made a deal with the two secretaries to just use one of the johns which included some old GE giveaways pocket size mirrors. I used the other johns to store files and brochures which I labeled, "combined-cycle, cogeneration, gas turbine, steam turbine, state of the art technology and upgrades" with the others used for "speeches" and their accompanying "slide trays" which could be used again. I not only liked my new "visiting" office but so did the cleaning woman. She only had to clean one john in there. She was so grateful she even cleaned the coffee pot I had in the waiting room. While she was going nuts trying to clean two buildings instead of three floors, she would take whatever time she could spare in there. She also knew I would help her with the overflowing trash bags which were now only carried out once a week.

I was ejected early from the "Work-Out" sessions by a Human Resources manager, Jan Smith, who was trying to become a Black Belt. The couple of times that Welch visited the plant, she was the one that went with the limo to pick him up at the airport and to plan his schedule. She had learned not to trust me when she asked me once what his favorite food was because I had told her I used to play bridge with the CEO Chef. She was going to have a cater supply food for a meeting in Genever-Watling's executive conference room. She had called the VP of Human Resources at headquarters but forgot to ask him about dessert. I told her he liked the decorative nature of watermelons stuffed with fruit balls, thinking that was an appropriate meal for the attendees.

In conducting her Work Out sessions, Smith often pointed to one of the new banner slogans developed from talks given by Welch which were often hung in the Work-Out meeting rooms: Some of these included: "Reality... Excellence and Ownership... Speed, Simplicity and Self-Confidence... Become Lean and Agile.... Bold Strokes... Need to Have, Not Nice to Have... Job Enrichment... Be No. 1 or 2 or you are off the train....Be a Bounty Hunter... Boundaryless.... Stretch.... Integrity is in Your Hands."

Smith was asking for more ideas to get rid of those "nice to have" versus "need to have" items to be more competitive. She had just commended a participant for his cost saving idea to turn the parking lot lights off at night because there was no third shift and it had saved several thousand dollars each year. Being aware of my friend, the cleaner, complaining about how she had fallen over one of the slow lane barriers trying to find her way in at night, I asked how Welch's new twin-engine helicopter was a "need to have" when he already had 25 private planes at his disposal at White Plains.

When I was told that our sessions were dedicated to Making Schenectady more competitive, I suggested employees could become more Boundaryless by being able go into those buildings before they were torn down to find some better furniture and equipment than they had to make their job more productive.

It was then I was told that my attendance was no longer necessary. However, it was Chuck Klages that got ejected a lot faster than I did.

In the beginning, she opened the Work Out sessions by reading from her 10-page newsletter entitled "Visions" which began with the announcement of the new Work Out program. It read: "The concept of Work-Out has a strong and direct link to our organizational structure, our organizational effectiveness and our operating plans. From an operational standpoint, the organization effectiveness program is diagnostic at the front end and action-oriented at the back end and functions on input from employees at every level in the business. From a business standpoint, the organizational effectiveness process is the bond which links our GE vision, its operating plans and key implementation strategies providing the basis for creating and managing change to improve the efficiency and effectiveness of the business as we go Boundaryless."

When Klages asked her to explain what the hell that meant, she held up Fortune magazine extolling Welch as the most admired CEO in the world. With great flourish, she declared: "We are seeing results in working capital turnover, operating margin gains and productivity gains, all adding up to net income growth. We need speed, simplicity and self-confidence in this vessel we have built to go into the rapids into the 90s -- not terrified by the choppiness of the ride, but exhilarated by the speed."

It was when Klages declared that he understood the simple-minded simplicity part that made her declare he was not a team player and that it would be best if he left and did not participate in the Work-Out program. He thanked her for "enabling him to make the company more competitive by not attending such bullshit meetings."

One would think Klages would have been fired. But he was one of those who took pride in doing a good job for customers no matter what was happening around him. He had 38 years of experience and knew every size turbine from flange to flange. Moreover, he did not have a college degree and was paid less than other parts specialists even though he was the one most of his counterparts went to when they had a question.

But as those attending the later Work-Out sessions said, one good thing did result from Klage's comments. Smith quit reading the mission statement from the Visions newsletter and just handed it out.

There was a time when Smith thought she was giving me good advice. She had been irritated when she came into my "broom closet" office to find it was overwhelmed by a huge leather chair that had once been used by the CEO Cordiner in the 50s when he had a visiting office.

She told me I looked ridiculous whirling around in it. She reminded me that the smart people were asserting themselves just like Welch wanted them to. I said shrewd, not smart, which I did not consider complimentary. She left when I asked her if she thought she was being assertive as a woman wearing such heavy perfume. Even Noll had kept his distance from her when she flipped his viewgraphs. I was to learn later he had an allergy.

Smith arrived at the Schenectady plant when Welch decided a few aircraft engineers should be sent to Schenectady to assist the gas turbine design team. Her husband Stan was one of them. The head of Gas Turbine had to let several good engineers go to make room for them. Smith was given a job in Human Resources. Soon she used her authority to stipulate that her daughter be given a summer job there. In the old days, lots of summer jobs went to college kids interested in engineering but those days were gone. Her daughter was the only one offered a summer job.

While "Work Out" with Smith as the facilitator provided lots of sick humor, Welch's Six Sigma program was not as funny. It really finished off any idea that our Schenectady plant could survive the drive for profits at any cost.

The program was billed as one of improving quality and delivery for customers. It was something all agreed was important -- but without investments in modern equipment, few improvements could be made.

So how did the Belts meet Welch's demands for zero-defects and faster deliveries under Sick Sigma? Often components had to be sent back by quality control inspectors for rework. By eliminating these inspectors, rework was eliminated and thus cycle time was reduced.

The idea not to use the new million dollar linear accelerator--which had been put in the Jones era of productivity gains through technology to spot defects in castings before expensive machining time and cost occurred --was devastating to customers as well as engineers. Shutting down the accelerator would do two things: save operating costs and by not looking for defects, then you would not find them. Six Sigma earned its Sick Sigma nickname big time when quality was improved by eliminating the people and the equipment that looked for defects.

Therefore, Six Sigma, Welch's Zero-defects aim was "achieved" at a great cost savings. Unsaid, but certainly known by even the dimmest of nitwits, was this would certainly boost the service business earnings. Unfortunately a lot of that work would go to those who bought many of the service shops during the "sell" mode by the Gold Collar VP of Customer Service anxious to boost his incentive compensation and stock options.

Work Out and Six Sigma meetings were also a way to weed out those who were not 'globally competitive" or "team players."

This time it was my friend Mark Markovitz who was ejected from a Six Sigma meeting when he noted that the only ideas accepted were those that saved production costs which often meant lower quality. He suggested the company could save even more costs by not wasting the time of manufacturing and engineering people who could be productively working on the product. Mark also got the attention of Genever-Watling when a large tent was set up in the huge parking lot near the Material and Processes Laboratory where he worked. Engineers were invited to come by with ideas to become more competitive. A huge helium-filled pig balloon with the names of competitors on it would be blown up by Genever-Watling's staff when enough ideas had been submitted to do damage to the competitors. Mark refused to join in the fun billed as Six Sigma "constructive conflict" by saying he would only do it if the clowns left the circus tent along with the pig full of hot air.

Welch, in walking his own talk, was to brag that of the 3000 Six Sigma projects in 1996 to 1997, productivity gains and profits of $320 million were achieved followed by $750 million in 1998 and $1.5 billion in 1999. There was never any hard data on these projects. Sick humor had it that the profits were really sales of technology and the businesses. As for investments he claimed to have made in the product departments, sick humor had it that most of this was the costs of tearing down buildings and moving offshore.

Most of the Gold Collar Black Belts held MBAs which to employees was often translated as "Move Business Away." Six Sigma was the end of the power generation business and not just in Schenectady. The eleven U.S. power generation, distribution and delivery business plants were reduced to four in the early 90s and by the end of the 90s they had been fully harvested, sold or moved offshore.

Only the small generator business, which was an integral part of the growing U.S. cogeneration business was left, because an offshore supplier could not be found to supply it cheaper. The only power generation action in the U.S. now was in South Carolina where the gas turbine business had been moved with the rest going offshore into co-production agreements or the technology sold outright.

During the Work Out, Six Sigma and Boundaryless campaigns, probably the most Machiavellian new subprogram was the one billed on banners as Job Enrichment which was, in effect, Job Consolidation. In the 90s, across the board layoffs -- along with the demand for actual names of those deemed in the Session C bottom rankings -- came at a faster pace. Sixty and 70 hour weeks became common for those trying to save the business. While doing this they had to stomach the slogan that their job had been enriched, not added to by unnecessary layoffs.

And it wasn't just employees being sacrificed for the bottom line under such slogans which were designed to sound ameliorating even though they were pejorative in implementation. Even the local assessors, who had gone through years of hearing the VP of Human Resources elaborate on the global goals of Boundaryless, Work-out and Six Sigma, had to figure out they were under a program called "Rationalization." Rationalization really meant tear down the buildings to get them off the tax rolls and sue for lower rates of assessment.

It was anything but rational to the city fathers who finally gave in to a whopping 75% reduction in reduced taxes and rates of assessment." Most damaging to the city was the demand to make these cuts retroactive. One assessor commented he could not understand why the Vice President of Human Resources thought the rate was so unfair. After all, the assessors had always taken GE's word for how valuable its plant and equipment were. Now they wanted to go back and lower the numbers they had given them in the past. Adding to the sick humor was the fact the costs of tearing down over 60 buildings were listed as "investments in plant modernization."

One of the sickest sub-programs of Boundaryless was called "Be a Bounty Hunter." The Gold Collar VP in charge of materials made himself several million dollars stock bonuses by offering monetary awards *(10% of the value of idea)* to those who could identify excess materials in the plant or old inventory which could be sold. The Green Belts had a field day scouring the plant looking for copper, steel and other valuable materials and parts. Manufacturing people went nuts trying to hide their inventory needed for production. Even some of the metals which had already been machined for particular parts were sold. There was even a case where a bounty hunter from marketing actually nixed a new order multi-million sale because he knew he could personally get a Bounty reward by convincing the customer to buy parts in inventory to upgrade his old turbine instead.

As for me, I tried to get put on the lack of work list when I saw that the heads of Engineering and Manufacturing I mostly served were being retired or laid off as each business was harvested and then sold, moved or closed. There were many of them who had done everything they could to save their businesses by resisting layoffs of technical people like George Cox *(the first casualty of the first Black Belt to be sent to the plant).*

I mostly remember Jack Hinchey, who had rebelled against Welch selling the aerospace business, including the Knolls Atomic Power Laboratory which built nuclear-powered submarines and was part of the power generation business.

His office overlooked the entrance to Bldg. 273. In the old days, it was considered the prime location. But because the Gold Collars did not want to be associated with the past they created plush offices at the front of the plant near Genever-Watling and later, Nardelli. Hinchey's office was also opposite the entrance to the balcony overlooking the 10-acres of machines below where my broom closet office was. Hinchey had often noticed me scurrying back and forth to my "visiting" office in the ladies room on that floor. He was also a weekend worker like Dave Wheeler and I were. Several times he expressed his astonishment at how jammed our office was with engineers as well as slides for them. On several occasions he also helped the cleaning lady and me haul out the heavy bags of trash.

I remember hearing him on the phone across the hall arguing loudly with Welch about the futility of telling the U.S. government they had to change the fixed rate of 12 cents earnings on each dollar to 15 cents. So when the entire aerospace business was sold, that was the end of Hinchey. Because he had been so important to the business, a retirement party was in order. But a manager being let go, even if they had been top management at one point, was also someone to avoid just as regular employees seem to avoid those being red-circled.

So I was named to organize his retirement party. I was getting all kinds of excuses from his peers they were busy that day. As was the custom, I called his wife to find out more about his personal life. When I discovered he was a friend of Tom Clancy, who wrote all those submarine books, I knew I was going to have a great bon voyage party for him. Clancy had actually written a book about Hinchey's experience as a submarine pilot during World War II. During an invasion into Japan, he escaped radar by settling the submarine on the bottom of the ocean. I spread the word Clancy was coming to the party. So I had a full house.

Hinchey thought the whole thing was funny. He even gave Wheeler and me a photo of a submarine for the invitation. But what really made us appreciate him was his suggestion (*and help one weekend*) that we move out of the old maintenance room closet and into his spacious office. All we had to do was trash the place with all of our photos, cutaways of various size turbines, post notes and rows of file cabinets. Then no one else would want it. He enjoyed our "decorating" his office that weekend as much as we did.

However, it was John Patterson who put up the biggest fight against harvesting, along with his head of gas turbine design Engineering, Don Brandt, and Leroy Tomlinson, who headed the Combined-Cycle business.

It was particularly difficult for Human Resources to replace them with Gold Collars because they were so popular with customers. Because the intention was to stay in the gas turbine business in South Carolina, Patterson had to be handled delicately. He was also very popular in the plant being expanded in Greenville. A solution would be to have one of his revered customers to hire him.

Because I had an inside source on what was happening in Human Resources, I found out they were tapping Patterson's telephone to see if he would accept any of these offers they were privately promoting. I will never forget my disgust *(and that of my informer)* when they called him the biggest fool they had ever known when he turned down an offer tripling his salary because he believed in the U.S. and did not want to leave the country. They simply had no shame -- and certainly no understanding of how keeping jobs in this country could be more important than their incentive compensation and stock options.

So they began rumors Patterson had an alcoholic problem. It was true he had a Martini at lunch. But it was vicious to suggest he was an alcoholic, particularly when they were having more than one themselves. Don Brandt and the rest of the engineering staff working for Patterson, of course, knew this was not true and they protested the rumors.

So Patterson did survive longer than most despite his refusal to the heavy layoffs they constantly demanded. Therefore Human Resources began sending out the letters urging early retirement if they were 60 or if under layoff notices to his engineers.

While Brandt did not have to deal with the Gold Collars as often as Patterson did, he was pretty good at putting them off politically by not arguing with them. Instead, he overwhelmed them with technical facts which caused them to leave him alone lest he find out they didn't know what he was talking about.

When our marketing communication was merged again – this time with a group that reported to headquarters – the job really became unbearable. I thought it was bad when my old boss (*yes, he had a family*) became a lobbyist in DC where they still remembered his father-in-law who had been a highly respected VP in the 60s. But he had done what he could to keep me from being my own worst enemy by letting me vent on him. Now, I was totally surrounded by Black and Green belts and those would-be belts who only saw their job as sucking up to bigger belts. Otherwise, they couldn't be bothered with doing any real work. This new group didn't even want me in their office area for fear of contamination so I found a meeting room four floors away from them and moved in it.

Sick humor was cathartic. And there were still some engineering managers putting up a good fight. I think it was LeRoy Tomlinson, who headed Combined-Cycle engineering, who really amazed me with his chutzpah. I had worked with him on the state-of-the-art technology seminars for several years. He was as precise as an engineer can get. His graphs of combined cycle operations were so detailed it would take weeks for graphic artists to get it right. We produced books of all the talks with the illustrations. It was in one of these that Tomlinson lost his temper. I was in charge of getting the books out as well as the speeches at the seminars. This was now part of my "enriched" job. He yelled at me that a pipe was missing in a schematic and another pipe now had hot and cold water running it in at the same time. I was yelling back that we missed putting the kitchen sink in that chart, too. He had made many changes on that one graph -- which was one of 50 other complex graphs just in his paper among 20 others in the entire book. I had not caught his last change when I gave the final o.k. on what seemed like a hundred proofs.

When he saw the area Dave and I were working out of trying to handle some 20 illustrated talks like his, he remembered that there had once been one whole floor of people in Bldg. 23 who did this in the old days and mistakes were still made then. He not only calmed down but he went to my Green Belt boss to say we needed more help. He came back amazed that he did not know what the state-of-the-art seminar was.

But it was Tomlinson's idea of perfection and not suffering fools which earned my greatest respect for him. The Tokyo Electric and Power Company (TEPCO) was installing the largest combined-cycle plant in the world. They were very familiar *(and appreciative of Tomlinson)* for his knowledge. When Genever-Watling was hosting the TEPCO CEO and his staff during a visit they asked to see Tomlinson. He called Patterson to get this guy there, whoever he was. Genever-Watling soon found out more about Tomlinson than he wanted to know. During the meeting, Tomlinson actually called Genever-Watling an idiot after one of his unknowledgeable comments about the operations of the plant. He was fired on the spot and quickly rehired when the TEPCO CEO said they would never do business again with GE again if Tomlinson was not going to be their engineering contact.

So with people with integrity like Patterson, Brandt and Tomlinson -- who cared about the business and not the stock option bonuses they could have gotten if they followed Welch's vision -- my job became satisfying for me

However, my Green Belt bosses were starting to demand seeing the talks I was doing for manufacturing and engineering so they could take out the best parts for the Gold Collars. It reminded me of the days when Lou Marsh would try to take all the good stuff out of my talks for Kurtz and put it in the talks of those he felt had the best chance of becoming CEO.

I began hating the job when I found myself working weekends for these Gold Collars whom Tomlinson had correctly proclaimed as idiots.

One weekend, in particular, sticks in my mind. A call came in from a Gold Collar Human Resource manager on a Friday afternoon about a meeting he had forgotten about until a secretary from the New York State Education Department asked his secretary if he wanted steak or lobster for his dinner entrée. It was then he remembered that he, along with human resources executives from Xerox and IBM, had been asked to make a two hour presentation each on what qualifications employees in the future would need to meet future technology needs.

At that time the plant phone system was still using forwarding identification *(when most calls on Friday were not answered and went to voice mail they realized the problem)*. I considered letting it go to voice mail but I was curious as to why he would call directly. Again, I was my worst enemy answering it. The next call I made was to my skiing friends that I would not be in Vermont that weekend.

Anyhow, after calling his secretary for a copy of the NYS program, I spent the weekend meshing my engineering and manufacturing talks into an hour long speech. They had recommended a 30 minute talk with audio-visuals *(two screens were available for slides and videos)*, and brochure handouts. This would be followed by an hour answering questions from the educators plus two hours in a wrap up panel session with the Xerox and IBM representatives. They were correct knowing that an hour was too long for a talk, but they did not know what I did -- that all this Human Resources Belt was really capable of was talking. I would fill up most of his question and answer period with videos, most of which I worked on back in my CAD/CAM days at headquarters.

When we got there *(we drove separately because I had not been asked if I wanted steak or lobster)*, I think he was amazed as I was of the huge audience. The 800-seat hall was nearly filled up with state educators. When we met the Xerox and IBM executives who spoke first, I knew we were in even more trouble because they knew what they were talking about. They easily handled questions of needed future skills. They also had a couple of staff members armed with several videos and boxes of brochures. They even had a projectionist with them although the state had provided one. I could see by the look on my Black Belt's face when he got up that he knew he was in trouble. Still, he read the speech with great flourish as I sat by the state projectionist to make sure the slides came on as marked *(I was so sleepy Sunday night I was not sure I had circled the right words for a slide change)* When he finished, my Gold Collar knew it was now "high noon at the corral."

What did he do? For the first time I heard myself being complimented by a Gold Collar. He announced I had some videos to show them and because I worked so closely with engineering and manufacturing, I could answer their questions and if not -- I certainly knew whom to put them into contact with. Then his ego did him in. He made the big mistake of telling the group he had to go because he had a very important meeting to attend. The looks that went around the room plus the look on both the IBM and Xerox executive was something I wish I had a video of.

But now it was high noon for me. Actually I enjoyed it. I probably got off into too much history of the Schenectady plant, but I managed to couch it in terms of changing technology. I even got a laugh when I noted that Edison had found a better substitute for the filament in his light bulb right here in the U.S. rather than that bamboo shoot he got in Japan which only burned for 30 seconds.

But when I was asked for copies of our brochures, I knew I was in trouble. I only managed to scrounge 30 copies each of our product lines brochures (*our global marketing manager had eliminated that budget and what I had was produced through engineering*). We needed them for customers. So to avoid the embarrassment of only having 30 brochures, I asked each classroom teacher to hold up their hand. I needed to get a count and then I would mail them brochures and other information they might want. I would pick up the addresses later. When no hands were raised, I was astounded. So were the guys from IBM and Xerox.

We had actually been talking to bureaucrats. I guess I became as insulting as my Black Belt human resource manager was. When one of the bureaucrats said they were the ones who planned the curriculum for the teachers and told them how to teach, I came back with my feeling that it was best to get the information to the people who were actually dealing with the students. Therefore, could I please have the addresses?

At this point, we were to have our two hour panel session up front. Now it was the executives from Xerox and IBM who, to my delight, said they also wanted the addresses of the teachers, who had direct contact with the students. During our panel discussion, they were quick to include me in discussing computer and electronic technology. I was thankful I had been with the CAD/CAM project set up by Jones.

When we were finished, I was about to leave when the guy from IBM asked why I was not staying for dinner. Even though he and the Xerox guy said they would insist I have dinner, too, I just laughed. I actually had to get back to work to finish a job for Patterson. So they skipped dinner, too. We went out for a drink at which time I congratulated them for not having a madman like Welch harvesting their businesses.

White collar employees like myself *(fortunately EEO and secretarial layoffs had gotten rid of the pink collar tag)* tried to keep our sanity with sick humor. One of our best laughs came when Welch's Human Resources Black and Green Belts plastered the plant with a Six Sigma poster headlined "Integrity is in Your Hands." This came after a Wall Street Journal article questioning fixed time cards in the Philadelphia switchgear plant plus a review of other charges of unethical and illegal behavior.

One would never know if Welch knew about these indiscretions or if it was his insane drive for earnings which caused his Gold Collar managers to take such risks. But whatever it was, the charges were occurring more often and the Journal *(it was not owned by Rupert Murdoch then)* actually acted as a corporate watchdog and ran editorials calling for corporate stewardship and responsibility.

So when the four-foot color posters, showing a pair of hands with a headline of "Integrity is in Your Hands" in huge type, were sent to each building for display, it didn't take engineers long to find a suitable place for their exposure. You could not go into one of their men's restrooms without finding the poster directly over the urinals.

If you frequented the bar next to Union Hall, which now had dwindled down to a membership of about 600 from the heydays of over 10,000, sick humor was also keeping the current survivors going. Before Welch became CEO, the Bldg. 273 bay areas did benefit from the installation of several CNC (*computer numerically controlled*) machines thanks to Jones investments to improve productivity. One particularly skilled union worker, who recognized me from the days of the GE News, told me he had worked his way up to operating the largest million dollar CNC lathe -- which could perform five functions which had previously had to be done on five other machines requiring costly setup times.

That was in 1980. But when Welch came in, he had been bumped down 14 times during the 80s and early 90s to lesser paying jobs until he was laid off in 1996. He went through a couple of layoffs and rehiring as older employees retired. He had two months to go to reach retirement age. His current job was to tear down five of the buildings he had worked in.

Then there was the electrician who had started out as a helper 40 years ago. His first job was to change the light bulbs in the huge 60-foot long Monogram on top of Bldg. 37. He worked himself up to being the lead electrician in the power plant. Large amounts of electricity were no longer needed so the power plant has been a casualty at the same time the foundries were taken down. He was recalled back to work for a week and found himself back where he started. This time he was changing the light bulbs for green and red ones because Genever-Watling's $7 million office was now under it. Unlike Noll, Genever-Watling did not deprive himself of "nice to have" items if they were in his surroundings.

Because of outsourcing, one had to be careful during customer tours. There was the time one of the manufacturing managers invited a group of utility customers from Georgia Power on a tour of Bldg. 273 after a meeting. Setting up tours was another "enriched" job I had. As usual, I proudly handed a steam turbine and generator brochure to each of them. Fortunately, I had learned to scout ahead of tours to warn foremen and the few workers on the floor we had customers coming through.

One of the factory workers pointed out a huge rotor forging that had recently arrived from Japan. It was even packaged in bamboo strips. Scrawled down the side of it were huge letters announcing it was for Georgia Power. I was able to divert the tour which earned me a dinner-for-two award from this manager, who was still putting up a good fight to save the business. Manufacturing and engineering managers were still able to give out dinner awards while Global old Collar Belts would often award expensive watches or golf clubs.

Having gotten so many dinner awards, I had congratulated myself on getting a Mexican restaurant added to the list. Many of my skiing friends also liked quesadillas.

During the 90s, Welch was constantly being hailed by Wall Street security analysts and the business press as the "Most Admired CEO." Naturally his Gold Collars agreed with them. Perhaps it was this champagne bubble he was living in that caused him to think his very words were enough to motivate the workers that were left. Among his many pithy statements to his adoring analysts was one in which he declared "employees are like a vegetable garden…they just need to be watered and fed."

This press statement was cut out and put on the plant bulletin board with a poem which was followed by many others. The one I remember most was "Jack Welch could eat no fat, Wall Street could eat no lean. So betwixt them, they licked the platter clean." Below that was Jack's eight figure salary and mega-millions in stock options. Another one went: "Little Jack Horner sat in the corner eating his profit pie. One for me and one for Wall Street." Another read: "Oh, he's huffing and puffing and he's blowing our house down."

These were quickly taken down by Human Resources people. But they did get their exercise finding all the time clocks in the plant where you could see their Six Sigma cutouts redone such as the one of a team of horses pulling together to turn the business around. It was revised to show just a couple of horses hauling a wagon full of Gold Collars.

But sick humor can only relieve the stress of insanity so long. Because my job had become so distasteful when I was pulled off real work to assist the destructive Gold Collars, I decided to walk out since I could not walk the talk. Besides, the Gold Collars had gotten rid of most of those trying to save the business. Only the yes-men and women were left and they were not only a dull bunch but lazy. And, yes, they had families so I should be more sympathetic.

My motivation to leave four years before I could retire obviously was not greed because I was going to take a big hit in my pension by just taking a vested one for my years of service. But I also could not stomach writing for making Black Belt Gold Collars look good nor did I want to join them at their resorts. I still had my health and there were plenty of places I still hadn't skied. I also like to swim, hike and was still not doing too bad in canoe and kayak races as well as in ski competitions. Most of all, I knew that I could now become a consultant myself because there were hundreds of laid-off engineers out there forming their own companies. They would need some help in marketing.

So out the door this "resource" went. I have to admit I tried for several years to get them to lay me off for the added benefits, unemployment and a month's pay for each year of service. But I had been my own worst enemy. I was still that girl that left the farm with a point to prove…that I could outwork anyone.

WHATEVER HAPPENED TO A FAIR DAY'S WORK FOR A FAIR DAY'S PAY?

That was the question I kept asking myself as I was going out the door at Schenectady GE. When I joined GE in 1960, we had our slogans, too. The company slogan, started back in the 1950s was "Progress is our Most Important Product."

Ronald Reagan, who was head of the GE Theater, kept touting progress is vital as he described GE's leadership in the consumer, industrial, power generation, transportation, military and aerospace markets.

This slogan was followed by "Accent on Value" which lasted ten years under the leadership of GE Board Chairman Gerald l. Phillippe and Fred J. Borch.

In 1972, GE Board Chairman Reginald Jones kicked off the "We Bring Good Things toLife" program. Under his reign, GE invested 10% of its profits into research and product development.

GE scientists and engineers were still getting the lion's share of industrial research awards. In fact, during the late seventies, GE technical people were receiving patents at the incredible rate of nearly four every working day.

Good things were also brought to everyone. Sure, we had our occasional programs to be more specific. For example, during an economic downturn in the mid-50s, CEO Ralph Cordiner came out with a program called "Operation Upturn: Build Sales and Jobs in 58." As an advocate of decentralization in the huge, diverse company, he reorganized the company into 120 different departments which were "not too big for one man to get his arms around." No longer was each business centrally controlled. Each department manager, most of them coming up through the ranks with a thorough knowledge of their product, was responsible for their business results. Downturns would come and go as did corporate programs to improve earnings.

In 1975, GE Board Chairman Fred Borch introduced "Best Buy" which not only boosted sales but also served as a motivational program. One in 10,000 employees received awards for outstanding contributions to the program. (*I am proud to say I was one of these.*)

Back then programs represented the company's desire to make employees proud and loyal – and customers pleased with GE products. The emphasis was on sales, jobs and earnings for investment and more product development. Those programs were always rewarding and fun.

But there was one time when I working at headquarters during the "We Bring Good Things to Life" campaign that banners were returned. One of the first batches was sent to the Burlington, Vermont plant where they produced Vulcan guns for the U.S. army. The plant manager, who sent the banners back, wryly noted perhaps Cordiner had been right. Decisions should be made at the plant level. Fortunately, I was not the one in charge of getting the press kits and banners to the plant.

But I did know the Burlington plant was proud of its work. In a companywide survey that same year to determine employee morale, the only business with a 100% job satisfaction was at the Burlington plant's Ethan Allen firing range. Jones, with a sense of humor, got a kick out of hearing that.

When Jones retired in 1980, GE was listed as the ninth most profitable company in the world. Some 450,000 U.S. employees were building products ranging from toasters to turbine.

But the slogans under the Welch era were demoralizing to employees, especially those in some 98 U.S. plants which were moved offshore under his three-circle program. We were battered with a never ending onslaught of slogans --Be No. 1 or 2 or you are out. Fix, Sell or Close; Bold Strokes; Need to Have, Not Nice To Have; Reality, Excellence, Ownership; Six Sigma; Work-Out; Boundaryless; Job Enrichment; Be a Bounty Hunter; Re-Invent Engineering and Manufacturing; Constructive Conflict (*an oxymoron*); Speed, Simplicity and Self-Confidence and finally the word "Stretch" was added which some thought might be a form of medieval torture.

As noted, in his 20-year reign starting in 1981, Welch transformed the company into a global financial business which was still big enough *(despite efforts to go back to technology by his successor)* to nearly bankrupt the company during the financial crises in 2008. Taxpayers came to the rescue along with Warren Buffet with a $10 billion investment to boost shareholder confidence.

As for those in the huge Schenectady plant, there are still many people today who remember the 30-month long campaign titled, "Make Schenectady Competitive." The company kept its promise to invest heavily in new equipment to achieve a 70% increase in productivity when the unions took pay reductions that averaged out at about 10% less than what they were making. Moreover, they have continued to see their wages stagnant ever since. The same has been true for many white collar employees. Everyone signed on to "a fair day's pay for a fair day's work" ethic to compete against offshore competitors.

Everything was going well and the union could no longer be blamed for making GE noncompetitive. However, when Welch became CEO in 1981, he got control of the company by introducing a new era of obscene pay, benefits and perks for the Gold Collars who had signed on to his GE Capital vision of its easier to make money on money. Employees became pawns in his financial gambles.

While Welch was dressing up his digs at headquarters and the management training center for $75 million "as a place for Gold Collars to bond," GE athletic centers in major plants across the country were the first "nice to have" facilities to be closed to save money.

The days when top management considered themselves well paid when they made 15 times more than individual employees were over. With IC, stock options and a seven figure salary, Welch alone was making 500 times, not 15, more than the average white and blue collar employee.

Prior to Welch, the GE average for earnings was 6 cents on the dollar with 10% of that plowed back into research and product development with the rest going to shareholders. Welch's demand of no less than 15 cents on the dollar *(with no investment in basic research and the closing of product development labs)* enriched a few for the short term at the expense of thousands for the long term.

Back when GE was paternalistic instead of materialistic, I remember when George Cox, -- after finding out an early turbine pioneer at the age of 101 years was about to be put in a county home because there had been no cost-of-living in his pension -- was able to prevent this by going back and rewarding him for patents which were still giving GE the competitive edge.

To see corporate materialism at its worst, all you have to do is google www.gmirating.com and read the report done by Researchers Paul Hodgson and Greg Ruel which shows Jack Welch at the top of their list of "Twenty-One U.S. CEOs with Golden Parachutes of More Than $100 Million." His total payout when he retired in 2002 was $417,361,902. Following him are Lee Raymond of Exxon Mobil at $320,599,861; William McGuire of United Health Group at $285,996,009; Edward Whitacre of AT&T at $230,048,463 and Bob Nardelli (*once head of the well-harvested GE power generation*) **of** Home Depot at $223,290,123.

And since I have started editorializing, excuse me while I do it on a broader scale.

It is time for politicians to put the country first and not act like greedy CEOs and bankers. Why should they be able to create their own lavish pay, benefit and retirement plans while the rest of us make do with a Social Security system we paid into which they now threaten to change by raising the retirement age.

The GE story under Welch is indicative of what is happening in the U.S. The rich get richer and the poor get poorer with fewer in the middle to support both. The rich have to ask themselves, do I really deserve the obscene money I am getting? How many mega-million homes on islands do I need and yachts to get to them? Can I just be happy never having to worry about money because I have so much of it now? Why not enjoy what I have instead of figuring out ways to make more of it and putting it in some offshore bank to escape taxes? Would my employees work harder if they felt pay levels were fair? If I were willing to work for a fair day's pay for a fair day's work like I asked others below me to do, would that be an inspiration to them?

What also galls me about these obscenely paid Gold Collars is they complain about not being able to find qualified factory workers and engineers. Factory workers and engineers who design and build products average pay from $25,000 to $85,000. Pay levels have fallen so low it is no wonder people seek jobs such as bus drivers, policemen, firemen and government workers *(fortunately for them their jobs cannot be moved offshore)* instead of working at GE and other engineering-manufacturing firms. A recent article reveals that a factory welder left his job at GE because he could make more as a maintenance worker at a local high school.

I am not against making money, just making huge sums by taking advantage of others.

There are those who have made obscene amounts who do a lot of good with it like Microsoft's Melinda and Bill Gates and Warren Buffet, who have given billions to offset global poverty and diseases. Even back in the robber baron days, people like Steel Magnate Carnegie established libraries across the country and Ford even started paying his workers a fair day's pay for a fair day's work so they could buy his cars.

But far too often the robber barons of today just hoard money they can't figure out what to do with. I have no way of knowing if people like Welch have ever contributed to society but I did hear once at headquarters that he had given a caddy on a Nantucket golf course $2000 to go to community college.

If it is not all about money, then it is ego. When one reads Welch's book, "*Straight from the Gut,*" it is amazing to hear the joy he got into doing deals, good and bad. Making it even more fun was he wasn't putting his own money at risk.

But from a woman's standpoint, some of the Gold Collar wives did benefit from the excesses. Welch's major personal expense was two divorces. The second one, who knew her way around legal circles, took a huge chunk (*$180 million*). GE Credit's Gary Wendt's wife ended up with $20 million from his $130 million estate. Both divorces got messy because all of the lavish perks they enjoyed were exposed.

I think the one detail that repulsed me the most was when Welch's Park Avenue apartment (*paid for by GE*) received a fresh bouquet of flowers each day even though he wasn't there. I could not help but think of Reg Jones who had ordered the weekly delivery of fresh asters to his office area be stopped and replaced with nice artificial ones.

Welch's first two wives deserved whatever they got as far as I am concerned --particularly the one who had to sit through all those dinners as Welch wrote his slogans on cocktail napkins.

Today, Welch is married to a former Harvard Business Review editor. Together they wrote a Dear Abby business column for Business Week with answers full of slogans and bumper sticker thinking. Most letters were signed anonymous or carrying foreign names. Even more amazing was the pandering Jack got from MSNBC and NBC (*owned by GE then*).

But his business show called "It's Everybody's Business with Jack and Suzy Welch" of advising officers of other corporations how to conduct business must have been too much for them. It was canceled after a month. The one I saw involved a pizza company asking for advice on expanding their markets. Welch and Suzy told them they needed to offer a menu other than pizza and suggested they have an employee in their spare time go out with a sandwich board to advertise the added selections. They were told that most of their shops just had two employees, a cook and cashier/waiter. The show was touted as a takeoff on Donald Trump's
"The Apprentice."

Welch and Trump are a lot alike. Both have outsized egos and no shame as they will do or say anything they can to stay in the public's attention. For Trump, it was his incessant questioning of Obama's citizenship even when the birth certificate was produced. For Welch, it was his tweeting that the U.S. Department of Labor was lying when it reported unemployment went down to 7.5%. It was amazing Welch had the gall to question it and say it was a political play considering he is the master of manipulators and never felt the need to back his data with facts. However, considering his record of laying off people, it is no wonder he thought unemployment was higher.

Someone handed me a brochure of one of Welch's ego trips – a special management program limited to 100 executives where you can even have your photo taken with him when you have lunch with him. I nearly threw up on my cocktail napkin when I read the front page of the slick brochure – "20 Years as head of General Electric, greatest manager of the 20th Century, bestselling author, business icon, and the world's most admired CEO – Jack Welch and YOU.

However, you can't fool all the people all the time. Bloomberg Business Week in their late December issue in 2012, revealed that when Jack opened his management institute back in January of 2010 as part of Chancellor University, (*a struggling for-profit college in Cleveland, Ohio*), the partnership did not last long. It was reported Chancellor and other for-profit colleges were recruiting students from homeless shelters and registering them so they could obtain federal student loans, which formed the bulk of the schools revenue. Not to belabor the point about people not being fooled all the time, the article went on to report "GE's uncanny ability to deliver steady earnings growth became less a sign of Welch's genius than his knack for moving money around and drawing on a richly funded pension plan."

The insanity of it all becomes very sad because most of the business press still refuses to look at the results of programs like Six Sigma under leaders like Jack Welch. However, one has to commend Fortune magazine, which twice hailed Welch as the most admired CEO and was now looking more closely at results, revealed that of the "58 large companies that have announced Six Sigma programs, 91 percent have trailed the S&P 500 since."

Even James McNerney, who served as head of aircraft engines under Welch and became CEO of 3M said introducing Six Sigma was a mistake because it "had the effect of stifling creativity so he removed it from the research function."

In fact, Economic Reporter Mark Truby, in an article in the *Detroit News*, says Ford Motor officials now acknowledge the company shoved aside many competent, experienced employees over the past few years in its zeal to identify leaders and fast-track future stars. CEO Nassar had emulated policies at GE under Jack Welch and now the automaker is studying ways to reward and retain technically proficient workers and reverse a loss of engineering talent that has contributed to lagging vehicle quality and troublesome new model launches. Said the new CEO: "We're back to getting professionals to do the job." So it is good news that the professional employee is once again being valued as more than a commodity.

On a personal level, my cousin who took an early retirement from ATK (*formerly Allegheny Ballistics*) which builds the weapons and surveillance systems for drones said he could not stand the amount of inane paperwork required in Six Sigma projects. It was introduced by some retired high military officers who often land in plush defense business jobs because of their political influence. He longed for the days of Admiral Rickover, who worked his way up to head the nuclear submarine business and was known for his safety record. When he became head of AEC, he had enough clout in Washington to ask that all political appointments be placed in one building with their fancy offices where they could pass paperwork among themselves without disrupting the business.

O.K. don't let me get started again– even though I was just beginning. On to more pleasant memories -- my new job as a consultant. Also, I was to have a new title having left the farm in West Virginia at the age of 17. No longer was I introduced as "Bill's daughter who never married." The pilot that put us in the lake was now Number One with me. We were married on Savage Island with just a handful of the "locals" there including 180 sheep roaming the island. And we spent a lot of time skiing at Alta and there was the time we almost....

Oh. Oh. I could digress again. So back to my job after GE. I became a marketing consultant. I was my own boss. No more 60 hour weeks. No more frustration with the insanity of Sick Sigma programs.

BEING MY OWN BOSS, WHAT FUN

Working as a consultant was less pay but a whole lot of fun. Most of all I did not have to suffer fools. Human Resources had not been able to fire or retire Patterson yet despite the vicious drinking rumors they were propagating. I formed a PR business called "Mountaineer Productions." Patterson immediately put me on the outsourcing list for manufacturing and engineering support.

Because outsourcing vendors had to join a new program called GE Partners, membership required them to follow GE's lead in reducing costs 10% each year. It was easier for me as a consultant just to take more time writing when these ridiculous across-the-board demands were made. It was not so easy for those retired or laid-off engineers who were hired by Customer Service to solve breakdown problems at customer sites. Their time could be easily checked.

Still, there were some who could give them the finger like Chuck Klages did. He was offered early retirement after his last run-in with Human Resources Smith. In his crowded Houdini office, he had pasted his politically incorrect pinup on the hard drive of his computer. As noted, he was one who knew turbines flange to flange and customers had not forgotten him. When a call came to him from a desperate Black Belt Gold Collar, he was going to go deer hunting. He said he would never come in for less than a $1000 a day. He told me later he had made up for all those years they underpaid him.

I continued to produce brochures for engineering as well as talks for Patterson at utility meetings. Again, I enjoyed the fact my talks could no longer be raided by the Gold Collars for their use. Patterson would just tell them he had done it.

Patterson was of great help when I became involved with those groups of engineers who had formed their own businesses such as retrofitting and upgrading turbines. A major problem for small startup groups was convincing the utility customers they could do the job in the time needed. After all, during a shutdown to check equipment or upgrade a part, a utility customer was losing thousands of dollars an hour in downtime.

One of the groups I worked with was called Innovative Controls. They started out as two engineers specializing in upgrading the Mark I, II and III Speedtronic controls of turbines. Ironically, they had been mistakenly laid off by Human Resources. Because Patterson had refused to name people for layoff, Human Resources -- not wanting the Welch team to know they could not control him -- had written in two names they found on the engineering payroll list. They had noticed that monthly checks were being sent to Indonesia and made the assumption that they must be some field engineers they had missed when they got rid of that group.

What they didn't realize when they sent out the usual form letters starting with ..."Due to the lack of work, we regret to inform you..." was that these guys were on site installing a new Speedtronic Control system for a major customer there. They were stunned when they got the letter. But Human Resources was not about to admit to a mistake. When the two engineers explained what was going on in GE, the customer paid them to finish the job. They came back to form their own business.

I was able to help. I got Patterson, who was well known with customers, to write a letter of recommendation for them as well as Arnie Loft, the developer and patent holder of Speedtronic. Both extoled their skills and dependability.

Loft and I had been friends back in the early 80s when he taught me how to use the Wang computer system. I had dubbed him the "engineer's engineer" when I helped him write his resume. Loft was headstrong and did not defer to MBAs, particularly the one that came in our food vending machine area one day.

I was using the microwave oven to pop a bag of corn I had in my purse. I had gone to the bathroom when the MBA came in and, hearing the popping noise, declared that the oven was going to explode. He rushed everyone out as the sound increased. It was then that I came back and declared that my popcorn would be ruined. Loft led the laughter at the Black Belt's embarrassment. When I was chastised for bringing in unsuitable food for a business office, Arnie also developed a taste for popcorn.

Loft was lucky because he, along with a lot of others, were close enough to retirement to go on pension and then form their own businesses as consultants. Loft was well known internationally as well as domestically so he had no problem naming his price. Others able to capitalize on his good GE reputations was Doug Todd, who had been quite involved with GE's Cool Water Coal Gasification plant in California. It had proven itself viable after five years of operation. But GE's participation in it was sold by Welch. Todd, with marketing as well as engineering background, continued to serve as a consultant on alternative energy projects.

Engineers with experience designing systems to burn waste were in big demand. One left for Oxford Energy, which was burning three canyons full of tires in Modesto, California and selling the electric generated into the grid at a substantial profit because they even got paid for taking the tires.

John Kovatich was known as "Mr. Cogeneration" in GE so he had no problem finding consultant work. U.S. manufacturers were looking for ways to offset energy costs by burning waste products from their businesses to produce electricity for their own use as well as for the grid. Other engineers joined Intermagnetics, a Latham New York company specializing in highfield superconducting materials. Its employees did very well when it was purchased by Royal Phillips Electronics.

Many engineers, either laid-off or leaving early realizing their jobs would soon be gone, joined Supermagnetics. This company in Schenectady completely revamped the old GE Maqua printing company building where in the early days I spent so much time on Thursday nights getting the GE News out. Supermagnetics specializes in high-temperature superconducting wire which provides enormous advantages over conventional conductors of electric power – high efficiency, smart grid compatible, green, clean, safe and secure. Even a greedy Welch would envy the money the engineers received when it was sold to Furukawa Electric Company.

Siemens established offices in Schenectady when it purchased Power Technologies, which was formed by laid-off and GE engineers who left because they did not like what was happening to the business.

But it was not as easy for those young engineers at Innovative Controls, who had to finance and establish themselves in a brand new company. We did produce brochures and the letters of recommendation from Patterson and Loft helped. While they could point to several well-done jobs to show customers they could get the job done on schedule, they were not well-known in the industry.

One day I took it upon myself to get their name out there when I noticed the Japanese were holding an international state-of-the-art (SOA) power generation technology meeting like GE used to do.

When I saw the program was filled with speakers from ABB, Alstom and Siemens, I noticed they did not have a session on upgrading gas and steam turbine controls. So I spent about an hour reading the papers on Speedtronic Control systems in the last State-of-the Art turbine technology book I had edited. I wrote a synopsis on upgrades for them to consider with a quick outline of all the benefits to customers. I submitted it under my own name, not wanting my Innovative Control friends to be disappointed if I did not succeed. The next thing I knew, my husband's fax machine was going crazy at two a.m. The Japanese were responding with a full program and contract for a breakout session on controls upgrade.

I delayed answering until I could tell the Innovative Control engineers, who were on a job in Brazil about it. By the time they got back (*and several nights two 2 a.m. faxes because of the time zone*), the Japanese had also offered to pay for the trip so it all worked out well. My husband, not enamored with my engineering skills, was relieved I was able to get my name off the program and substitute theirs. But most of all he was glad not to be woken up when his fax machine ran out of paper. However, I cherish my name on the first program listing me as an expert on Speedtronic Control systems upgrades.

Not having lost my penchant for sick humor, my best job was managing the GE Hall of Electrical History for the Elfun Society. I had not been a fan of the Elfun Society because in the old days you had to be a high-level GE employee to belong. When I was at headquarters, (*EEO and Civil Rights legislations going into effect*), I got a letter along with a black employee, Willie Campbell, inviting us to join this "influential group of company professionals, business and technical leaders."

When I got to the exclusive Fairfield County Chapter, I was asked by the host, who assumed I was there to work, to collect dues from Elfun members who had not paid for the dinner. I thought this was funny and had a grand time fulfilling his request until my VP came in and disgustedly told me to go to the table for new inductees. In about 10 minutes, I was joined by Willie, who had been told by this same host he was late for his caddy duty. We had a big laugh over it but not as big as Mark Russell. In those days he was a well-known piano playing pundit, who was paid handsomely for his political entertainment. When he spotted Willie and me, he just wouldn't let it go. After learning our names he quickly incorporated us in his politically incorrect repartee at the piano questioning what we were doing there.

When I left headquarters and was back at the Schenectady plant I got a letter telling me I was no longer eligible for membership because my corporate pay level had dropped. Therefore, when Welch, as part of his plan to set up management according to his vision, got rid of the caste system in this organization and opened it up to everyone, I must say I was pleased.

So here I was running the Elfun Hall of History. It paid $10,000 a year and supposedly I was only to work two or three days a week. But I enjoyed it so much I usually ended up there a full week.

I was into GE history big time -- particularly Charles Proteus Steinmetz -- a brilliant mathematician known for his laws of hysteresis which put Schenectady on the map as the city that lights and hauls the world. While Edison and his Menlo Park inventors came up with the light bulb, it was Steinmetz who wanted everyone to have one. He was able to prove electricity could be distributed long distances with cheap hydropower at Niagara Falls. Still, I was glad the nation's honeymooners objected to him damning up the whole thing for hydropower generation. But his heart was in the right place. As a socialist, he wanted the best for everybody and if everyone was honest like him, it would be a great system.

I still quote a fifty-year employee I interviewed when I was doing the Pensioner's Page in the GE News. He knew Steinmetz back in the early 20s. He told me how Steinmetz had come to this country seeking out Edison, whom he admired for his inventions. As a hunchback, Steinmetz looked even worse for his two-weeks in steerage coming over. When the back of his jacket was marked to have him sent back as unfit, a passenger who knew of his genius vouched for his ability to take care of himself. Steinmetz immediately sought out Edison but was rebuffed because no one could understand his guttural German -- except for another German, Dr. Eickemeyer. He ran a hat making business in Yonkers and hired Steinmetz. Soon the International Electrical Engineers became interested in Steinmetz who started writing about his mathematical theories. Dr. Eickemeyer was one of the first beneficiaries when his assembly line was no longer dotted with pulleys driven by steam power, but motors that did not burn out.

The loyal Steinmetz refused to leave his benefactor despite Edison's pleas. So when J. P. Morgan -- who was trying to solve the problem of patent infringements holding up the electrification of America – began merging companies (*including the Edison Works into GE*), he simply bought the hat maker's business to get Steinmetz.

Completely enthralled, I asked the old timer what he felt the difference was between Steinmetz and Edison since they crossed paths in Schenectady and he had met them both. He said, "It is simple. Look around Schenectady. The parks and schools are named after Steinmetz and the country club is named after Edison."

The old timer was one of those thousands of Italian and Polish immigrants who came to Schenectady to work in the shops during the early 1900s. There were not enough schools to accommodate their families so Steinmetz ran for the School Board. His campaign was simple, "A Seat for Every Child."

The 50-year-GE veteran also told me Henry Ford had once come to Schenectady to seek out Steinmetz, who was known by the engineers as "the Supreme Court." He wanted some advice (*people knew Steinmetz was free with it*) on putting headlights on his Model T's. Steinmetz solved his problem but not before asking Ford to set up schools for the newly-arriving immigrants in Detroit which he did. When Steinmetz died, the Ford Museum moved Steinmetz' Mohawk River Camp -- where he studied the forces of lightning by observing its path in a broken mirror -- to Dearborn.

But I digress -- it is easy to do when it comes to Steinmetz. Anyhow, maybe it was his Robin Hood attitude or my deep regard for preserving the history of the better side of GE, but whatever it was, I thoroughly enjoyed leveraging the assets of the Hall of Electrical History with some bargaining of my own. During the Welch era in the 80s and 90s, the Hall was always just ahead of the wrecking ball. The Elfuns were constantly looking for space in basements or warehouses no longer used for inventory or buildings being vacated because of move outs and layoffs. Our last building before we were told we had to leave the plant was the one Steinmetz had conducted his artificial lightning machine experiments in to show how insulators and other devices could assure safer electrical transmission systems. A five story building, it had been built with reinforced steel bars in concrete. It was probably the most costly to take down.

We preserved over a million photographs from the Edison and Steinmetz eras plus hundreds of artifacts such as the early Edison bulbs, the first toaster, fan and even an electric cigar lighter. But it was the comprehensive documents from the past, many of them rare and forgotten that proved to be most valuable in keeping the Hall of Electrical History financially solvent. I must admit I particularly enjoyed "leveraging" a couple of sales to headquarters.

When a call came in from the receptionist at the White Plains airport that GE wanted to acquire a string of old time Christmas bubble lights as a present to Jack Welch, I was about to laugh. But I quickly squelched it into a most accommodating service manner to find out that as a boy Welch had loved these on the family Christmas tree. The receptionist told me that it was to be a surprise to him, that she was going to put them on a little tree and place them in his twin engine helicopter.

I told her how rare it was to find one of the early light sets, particularly in working condition -- and that we did have one but had even turned down a major Japanese customer steam turbine customer rather than de-accession it. She reminded me that Jack Welch was the CEO of GE and I noted that even so the Japanese customer had offered us $2500 for it. As soon as the check for $3000 arrived, the bubble light set which I bought at a garage sale for $2 was on its way.

Since that was such an easy sale for the Hall, I searched our files for photos of GE powered planes which I could copy for sale. Among many, I also came up with the first GE mail plane being christened by an attractive blonde wielding a bottle of champagne. I described the woman knowing full well I was describing the attractive White Plains receptionist because I had been there many times. I got $1800 for a framed photo of that one since it was the only one known to exist, having been shot in Schenectady. Other framed photos of the first jets powered with GE engines garnered from $600 to $800 each. Even well-connected receptionists feel the need to bargain.

But undoubtedly the biggest payoff for the Hall of Electrical Industry came from the documents which included papers and even an ad expounding the use of PCBs in transformers. While

GE's defense had been they did not know the dangers of these chemicals until the ban in 1972, there were low-level engineering papers noting such harmful effects much earlier. It was tough standing up to some GE lawyers intent on this information not getting out. However, they did agree to some donations to preserve history as long as it was not about PCBs.

Another big donation came from a phone call about an old Hartford GE lighting plant which had become a residential building. The tenants claimed to be suffering from asbestos fibers and a class action suit was being made against all the owners of the building over the past 100 years. Did we have any information about the building when it was a GE plant? It took several hours but I found it including photos of the plant. The lawyers, having dealt with me before, assumed I had found information about asbestos. Another large donation was made.

I was really beginning to enjoy the job when the Hall was notified it had to leave the plant. There was no empty space left, including an old classic Edison Era building we had tried to move into. It was torn down to prove the point that no one should rest their laurels on the past, even the Hall of History. Despite a treasury of over $200,000 in just 18 months, we did not have enough money to buy or build in the city.

We agreed to merge with the Schenectady Museum, which turned out to be a mistake. It is a city museum without a focus. Before we left I had managed to convince my old nemesis, Smith, now head of Human Resources, that the Hall should receive $20,000 for conserving companywide photos and information and updating GE history. When I noted that she would be a part of GE history as well as Welch since the Hall updated the GE history book every 10 years, she laughed.

I couldn't help but remind her about a letter Mark Markowitz had sent to the Editor of the Albany Times-Union about her receiving the Albany Chamber of Commerce award for being an outstanding business woman. *(Many organizations had learned that if they wanted a grant from the GE Foundation, it would behoove them to complement Smith).* Markovitz, pointing out Smith had overseen the elimination of thousands of jobs, wondered why the Chamber was not giving awards to women who started new businesses and hired employees instead of those going out of business.

Thus, when I told Smith the $20,000 would be needed because I would be answering the Hall of Electrical History phones at the Museum – and that one could never know what I may have to say when GE customers called in for a photo or information -- she used Welch's four letter words for a while *(shocking the old guard GE trustees)*, but agreed to it.

Unfortunately, the not-for-profit professional museum executive director there was told by her the money was for actual photos and information that was requested. He did not question this and thus they lost the potential I had found in "realizing" the value of the archives.

So while the history is still there – in a temperature controlled atmosphere thanks to our $200,000 – my big regret was not checking out the financial status of some of our members, particularly Bill Foote. He died leaving $5 million, which only took this not-for-profit museum ten years to spend. If we had known that kind of money was coming into the Hall of Electrical History, we would not have merged with the Schenectady Museum.

I left soon after the merger because dealing with bureaucracy up there was no fun. The executive director there accused me of being too business oriented and reminded me that we were a museum, not a profit center. He declared I must think I was Jack Welch. I couldn't stop laughing. It would take too long to explain myself to him.

What I did was leave him with several boxes full of Welch banners.

So what am I doing now besides bossing myself? Now I am just one of those retirees who do not receive automatic cost of living increases in their pensions. Fortunately, what we contributed to Social Security does have c-o-l adjustments so many of us have seen that surpass our annual GE pensions.

If you have gone to retirement parties then you know it is these parties which let you know if the person was respected or not. The one I remember the most was one for a manager back in the old days before computers. He would not read a press release or speech unless it was typed perfectly. Worse, he often made changes three or four times. It would take days to get him to approve something and this included time spent outside his office waiting for him to see you. He was speaking at the men-only Lower Mohawk Club in Schenectady when I was told to type in a new thought he had in his speech and deliver it right before lunch at the club.

Despite my protests when I got to the door with the clean speech, I was rudely shoved out by two men. I sat down on the steps and was lighting a Virginia Slim cigarette (*we obviously still had a long way to go baby*) when General Managers Alan Howard and Fred Glockner showed up for lunch. They were disgusted I had been thrown out and wanted to sponsor me for membership which I quickly declined knowing it was half my salary. However, because I lived at the end of a driveway above a garage near the club which was often mistaken for the club entry, I put up a sign that read, "Everybody can turn around here except Mohawk Club members." This got me a letter notifying me that I now had sitting room privileges at the club.

But I digress again. Back to the clean copy speech giver. Because he was a high-level manager when he retired, a huge obligatory party was held for him. I was told to go along with a photographer to prepare a report on the festivities. Just as soon as dessert was served and his boss announced he was now officially retired, everyone left except the manager's family. They look so confused I felt sorry for them. I decided not to use the picture we had taken of the empty room in his memoir retirement book.

Anyhow, by walking out early I did not have a final GE retirement party except for one given to me by my skiing, canoeing and bridge friends. But I did have a bon voyage party from headquarters twice. One was when I left Corporate Employee Relations. Among the gifts *(people had to contribute since I was not high enough level to receive a company-paid one)* I did get a gift certificate to Macy's plus a hoe which I laughed about thinking it was another referral to my West Virginia background. Instead it was presented as a preparation for me should I ever go back to upstate New York.

But the bon voyage I enjoyed the most was the one from Kurtz and our CAD/CAM group. The 100-year-old Ballantine was flowing freely. So when given a Gucci briefcase by Kurtz, which I recognized was worth several thousand dollars, I asked him for the receipt. He was still laughing when the party ended.

So even if I don't make it to 100, I can make it with Social Security, my vested GE Pension and investment income plus a house in West Virginia where taxes are low because we do not have four levels of government, just three county commissioners, a sheriff, a county clerk and one school system. So while I am lucky-- that doesn't stop me from wondering about what the future holds for people looking for jobs today. I also think about some of my former GE friends who were in their fifties when laid off or had their pay red-circled down to entry levels. And yes, I do have some GE afterthoughts – particularly now that Welch is gone and his replacement appears to be another Reg Jones. That is next.

GE TODAY, GOOD LUCK IMMELT

Today when a stranger passes by the Schenectady plant, he or she might be impressed by the landscape dotted with green lawns and a few buildings surrounded by an artificial landscape of trees and shrubbery.

They would never suspect these same grounds once boasted over 300 buildings with a workforce that averaged 25,000 during the 60s and 70s (*during World War II it was over 40,000*). They probably would not realize that as a taxpayer they helped pay for the landscaping. After the plant was harvested by Welch, his PR team got a government "industrial park" grant to spruce it up.

Hopefully the last motor built there, called the Phoenix, is an omen for the future. When Jeff Immelt came in as CEO in September, 2001 one of the first things he did was restore the technology growth rules of putting 10% of profits back into basic research and product development. Just as important, he got rid of the three-circle strategy with its "fix, sell or close" mandates which not only tipped off competitors to wait for the sale and customers to flee but was so demoralizing to employees.

Immelt is learning how much harder it is to grow a business rather than harvest one. And it has not been an easy job. Under Welch's twenty-year reign, factories were not modernized so it is almost like starting over to catch up.

Even worse, Welch left Immelt with a $9.4 billion bill because his "financial engineering" resulted in him underfunding GE Credit's insurance business that amount -- allowing him and his cronies to walk away with additional millions in stock options based on false earnings.

As Tom O'Boyle stated in a Bloomberg Business Week article, "you can't evaluate a CEO's legacy in the time he was CEO. You have to look at what was laid at the successor's feet. And on that criteria, the market cap is less than half of what it was when he left. Doesn't that somehow count toward the consideration of what he did while he was CEO?"

An early answerer of this question was Economist John Cassidy in a New Yorker magazine article right after Welch retired, titled "How great was Jack Welch?" Noting Immelt will be hard pressed to maintain GE's earnings growth rate because it was dependent on GE Capital and internal banks that were in the forefront of Welch's acquisition binge, Cassidy stated: "Welch's draconian policies certainly boosted profitability, but the numbers are not quite what they seem."

Cassidy pointed out research and development was cut and no revolutionary products developed which led to growth in the past as he reviewed GE's financial engineering under Welch. Here's how he explained it: "Every year, Welch bought dozens of companies to boost GE profits and its stock price. The process worked like this: Say that GE has a stock market valuation of four hundred billion dollars and profits of ten billion dollars, which means the stock is trading at forty times its earnings. Now assume that GE buys another company with a stock market valuation of twenty billion dollars and annual profits (*or potential profits*) of two billion dollars. What is the value of the combined company? You might think the answer should be four hundred and twenty billion, but that's not how Wall Street sees it. If investors continue to believe that GE is worth forty times earnings, the new valuation will be four hundred and eighty billion dollars. As if by magic, sixty billion dollars will be created. In relying on acquisitions, Welch exploited the myopic way in which Wall Street often values companies. As long as GE delivered consistent earnings growth, which it managed to do for 20 years, the strategy has been sustainable. But truly great businessmen such as Thomas Watson, the creator of IBM, do not rely on financial trickery. They built durable business franchises that last for decades. Welch did no such thing."

And no one knows this better than the tens of thousands of employees who did not have to lose their jobs in the process of GE becoming a financial business. Plants like Schenectady were making money, just not enough to please Welch. If they were not starved of re-investment dollars for modern equipment, many of these businesses would be around.

Not only are the days gone when Schenectady GE had the talent and modern plant and equipment to build the products to light and haul the world, but it takes years to develop the talent and make the investments to once again be the world leader in power generation.

But true to GE's history of innovation, Immelt is taking the plant in new power generation directions. With climate change and the need to rid the U.S. dependence on Middle East oil, the company is taking a serious look at producing "green" energy.

Schenectady employment is now climbing with new jobs for the first time since 2003 in alternative energy such as wind generators. In fact, Immelt has visited the plant several times to encourage the troops. His message is of hope for the American worker, when he notes that "American workers can hold their own against offshore competitors because they are not only capable but highly competitive if given the chance." The chance, of course, is an investment in new equipment.

To bring the company back to its technological roots, Immelt seems to realize, as Jones did, that it is modern equipment which enables productivity improvements of up to 70% -- not being fearful of losing your job. With wages now commensurate with most global companies, the most reliable source of earnings continues to come from producing a product -- not taking the high risk of making money on money.

In fact, in 2011 Immelt hosted a visit by President Obama who toured the plant, praising workers as a model of what is possible for economic growth through exports and smart planning. He commended GE for building a battery plant and locating its headquarters for renewable energies such as wind power at the Schenectady plant.

This investment in the Schenectady GE plant came with an $100 million investment by GE in conjunction with a $25.5 million tax credit from the federal stimulus package, $15 million from the state and $4 million from the local Metroplex Authority and other tax incentives. The union also took more heavy cuts in wages to attract business in renewable energy products and services.

While a lot of taxpayer money did go into the Schenectady plant, it is not even close to the deal that Global Foundries in nearby Saratoga County is getting for building a $4 billion computer chip plant. Local school, property, sales and state tax breaks along with an outright investment of $1.2 billion by the state has resulted in a taxpayer subsidizing of over $1,200,000 per employee.

In a press release, an alternative energy manager said, "Success for us is to turn this into GE's next billion dollar business. But for the long term, GE needs to learn how to build a company from scratch." To me, it appears this is a manager who must remember how GE product businesses were harvested instead of nurtured.

In May, 2015 GE announced the launch of its Digital Wind Farm, an adaptable wind energy system that pairs turbines with the digital infrastructure for the wind industry. The technology boosts a wind farm's energy production up to 20 percent. The digital wind farm uses interconnected digital technology, often referred to as the industrial internet, to address a long-standing need for greater flexibility in renewable power. It will help integrate renewable power into the existing grid more effectively. The system begins with the production of the turbines themselves. With the next generation of wind turbines, GE's new 2-megawatt platform utilizes a digital twin modeling system to build up to 20 different turbine configurations at every unique pad location across a wind farm in order to generate power at peak efficiency based on the surrounding environment. Additionally each turbine will be connected to advance networks that can analyze turbines operations in real time and make adjustments to boost operating efficiencies.

GE has invested heavily in solar and wind because these forms of energy employ GE equipment such as wind turbines and power inverters. The company has invested about $10 billion in 17 gigawatts of renewable power since 2006, along with more than $1 billion in clean energy projects such as wind and solar.

GE owns part of the 550-megawatt Desert Sunlight Solar Farm which is being built using GE power inverters. Wind farms under construction or completed across the U.S. and in other countries like Ireland use more than 4400 GE wind turbines.

In 2014, GE expanded its renewable energy horizon by making its first investment in a solar power project in India -- $24 million in an 151 megawatt solar array project.

Sometimes it takes years for an investment to pay off big time and the fuel cell business may be one of these. It took today's hugely successful gas turbine business over 10 years before it made a profit. Seventeen employees are now working in a facility near the huge Global Foundries computer chip business to develop fuel cells which can power large industrial data centers, entire neighborhoods and even developing countries.

To bring the company back to its technological roots, Immelt has also invested $500 million to formulate the technology for a 60 Cycle Combined Cycle plant using a 7F7 series gas turbine and an H-26 generator and $170 million on test facilities to validate it. While the gas turbines will be built in South Carolina and offshore, GE Schenectady will once again have a profitable generator business in Bldg. 273, albeit a small one -- not like the ones GE used to build for steam turbines rated up to 1200 megawatts, a business Welch harvested because it wasn't making as much profit as he wanted.

The tremendous gains in combined-cycle fuel efficiency along with the ability to rapidly increase or decrease power output in response to fluctuations in wind and solar power, promise a bright future for the 60 Cycle Combined Cycle plant.

Announced in late 2013 was an order for three contracts worth $2.7 billion to build 26 gas turbines, 12 steam turbines and 38 generators for Songelgaz, Algeria's state-run electricity and gas company. Twenty-four simple-cycle gas turbines were shipped in 2013 to this same Algerian customer, adding about 370 megawatts to their grid.

In March, 2015 GE announced a $1.9 billion deal to supply the Egyptian government with 46 gas turbines to create 2683 megawatts of electricity.

The work is being split between Schenectady GE and Greenville, S.C. It is providing jobs for some 1150 factory employees in Schenectady GE's Power group through 2015. Over 4000 employees now work in Schenectady GE's main plant after going down to under 1500 when Welch left office. Still, old timers wish for the old days when employment averaged 25,000 before Welch starved the power generation business to turn the company into a financial conglomerate which proved to be disastrous. They remember when a single steam turbine or hydrogenerator system could produce over 1000 megawatts of electrical energy. Today, German and

Chinese workers are thriving building those huge steam turbine and generator units.

But GE's largest acquisition in 2015 – its purchase of France's Alstom Power for over $10 billion – should put us back in the power generation business big time. This deal includes three joint-ventures in grid technology, renewable energy and global nuclear and steam power. This deal may one day result in GE becoming the major power generation player it was before Welch harvested it.

Ironically, it was Welch, who sold GE's advanced gas turbine technology to Alstom back in the 1980s. Now we have it back. Most important, the deal also gets GE back in the large steam turbine business which was so totally harvested by Welch that Siemen did not buy it as predicted by the business press. So GE simply went out of the large steam turbine business. Now it is up to Immelt and his successors to reseed the garden Jack harvested.

The purchase of Alstom is a great start and Schenectady once again becomes the headquarters for a broad array of power generation and energy delivery products.

Now called GE Power, it is headed by Steve Bolze, president and CEO, and contains six business units. They are Distributed Power, Nuclear Energy, Power Generation Products, Power Generation Services, Renewable Energy and Water and Process Technologies.

Bolze, in announcing that the purchase of Alstom now makes GE Power the largest industrial division in the company, said plans call for focusing efforts on the growing energy demand across the globe. "As we continue around the world, Schenectady is our headquarters," he said, "and it will continue to be a showcase and that is why we invested here."

GE is also investing $15 billion in the fracking business which includes a laboratory to develop cutting edge science to improve profits for clients as well as reduce the environmental and health effects of the oil and gas drilling boom.

Immelt also invested heavily in becoming the supplier for energy infrastructure equipment by buying the well support division of John Wood PLC, which makes submersible electric pumps and Wellstream PLC, which makes flexible subsea risers that connect wells to the surface and pipelines. GE also bought Dresser, Inc. which makes valves and controls for oil production. In 2013, GE paid $3.3 billion for Lufkin Industries, a drilling-equipment maker for shale drilling.

Profits from these investments paid off until the price of oil went down, slowing the gas drilling business. While Wall Street analysts give Immelt a mixed rating on getting into the volatile oil exploration business, others say Immelt did the right thing and at a certain point in time the market is going to appreciate what he's accomplished.

Another business with great potential is the GE Healthcare business which is using 3-D imaging technology first developed at the Global Research and Development Center. After collaboration with the Massachusetts General Hospital, it has been approved by the FDA. The technology reduces blur and increases the sharpness of an image.

GE is now the largest producer of diesel-electric locomotives for both freight and passenger applications in North America, reaching a 70% market share in 2015. The main manufacturing facility is located in Erie, Pennsylvania. To meet increasing sales, a second locomotive factory went into operation in Fort Worth, Texas in 2012. It is refreshing to see that Immelt expanded this business in the U.S., not out of the country.

The current GE Evolution Series Locomotive is the result of a 10-year, $400 million investment by GE Transportation introduced in 2005. Today, more than 5000 of these locomotives are operating in ten countries. GE's latest new hybrid locomotive can move a ton of freight 500 miles on a gallon of fuel. The innovations for this heavy haul locomotive came out of Immelt's heavy investments in research and development.

Thanks to Immelt, GE is now a major leader in aircraft engines. In 2013, GE and its partners, CFM International (GE and Safran, a French aerospace company, and the Engine Alliance (GE and Pratt & Whitney) won more than $40 billion in commitments for advanced, fuel efficient jet engines. GE's Leap technology -- which features low weight carbon composites to save fuel and heat resistant ceramics – was also developed at GE's Global Research and Development center in Schenectady. GE's revenue from the aviation business is now 14.8% of its total.

In fact, just a $75 million investment in GE's Rutland, VT, aviation plant to shape advanced materials such as titanium aluminide has led to more than $300 million in engine production savings while creating jobs. (*That investment sum reminds me of the same sum Welch spent on headquarters amenities in the 80s.*)

While I do not believe in corporate welfare, I must say the free tax rides Immelt has gotten are more justifiable than those of Welch. Under Welch, GE did not pay corporate income taxes for three years because his lobbyists were manipulating the systems with tax loopholes. When Welch left GE in a financial mess, it was ironic that GE under Immelt also had three years of not paying taxes because it was able to claim the heavy losses of GE Capital.

Since the U.S. financial crisis in 2008, GE has been steadily paring back its finance arm to recover from the disaster. As Peter Eavis in the New York Times explained, "the financial crisis delivered GE a near death blow. Struggling under a portfolio laden with risky loans, GE was bailed out to the tune of $139 million in government-guaranteed debt and lost its prized AAA rating."

Warren Buffett also came to GE's rescue with a $3 billion purchase of preferred GE stock which he earns 10% on -- which I have to say is about triple of what us regular shareholders get. Still, I have to admire Buffett for noting that his secretary pays a higher income tax rate than he does. He even called for the rich like himself to start paying their fair share.

So Immelt's announcement in April 2014 that GE will sell most of its GE Capital assets by 2018 to reshape the company and further the role of its industrial business is welcome news for those seeking jobs. Immelt said that by 2018 GE's industrial businesses will generate at least 90 percent of GE's operating earnings. GE Capital assets targeted for disposition, in addition to GE Real Estate, include most of the commercial lending and leasing segment and all U.S. and international banking assets. GE will retain those financing businesses that directly relate to its core businesses such as aviation and health care.

In 2015, GE invested $5.5 billion in the Global R&D center which covers everything from undersea oil facilities to bioengineering to aviation.

So hopes are high for GE as Immelt brings back reinvesting in research and product development which has taken years to pay off because he essentially had to start over.

In fact, just from 2003 until 2013, Immelt invested $43 billion in research and development, 35,800 patents have been filed and 45,000 engineers and 50,000 sales and service people have been hired. So he has made a great start in getting GE back into the business of growth through investments.

After all, it took Edison and other new product developers following him to make GE a profitable company for over 100 years. When Jones turned the reins over to Welch, GE was still taking the honors in getting the most patents. It is good to see us back in the business of developing technology and getting patents.

As Immelt told Leslie Stahl on TV's 60 Minutes, "the mistake we make is by not making enough bets in markets that we're experts in."

By putting great emphasis on research and development to develop everything from enhanced breast cancer and brain imaging equipment to more efficient turbines, aircraft engines and locomotives and solar energy research, he has revived the Jones era of "Progress is our most important product" and "We bring good things to life."

Pleased to see that Immelt is bringing the company back to its technological heritage which also creates jobs, one remembers when Welch made one of his pithy condescending comments that "employees are like a vegetable garden, you need to feed and water them." Reality has shown that he not only got rid of over 180,000 middle class jobs, but he was very successful at harvesting the garden by not reinvesting profits in research and product development and selling off advanced technology and product businesses.

But now that Immelt is bringing back jobs in durable businesses, workers no longer rely on sick humor to deal with Jack and his Machiavellian slogans and programs like Sick Six Sigma. The plant now has an atmosphere of working hard for a cause. Loyalty is replacing fear as a job motivator.

Jack's style of using and abusing Six Sigma to harvest the business, which was copied by other greedy CEOs, is now being revealed as harmful to other major U.S. businesses. One of the most vivid examples occurred at 3M, a company that prided itself on drawing at least one third of sales from products released in the past five years. In an article entitled "At 3M, a Struggle between Efficiency and Creativity," Business Week's Brian Hindo outlines the Six Sigma approach brought to that company by James McNerney, a Gold Collar Six Sigma head of GE's Aircraft Engine, who was a major contender to succeed Welch. As Brian Hindo points outs, "McNerney had barely stepped off the plane before he announced he would change the DNA of the place. His playbook was vintage GE. McNerney axed 8000 workers."

Just as Welch did at GE, McNerney intensified the performance review process and trained thousands as Six Sigma Black and Green Belts. No new products were introduced during his four years at the helm. During his reign he slashed capital expenditures, including huge reductions in research and development from 6.1% as a percentage of sales to 3.7% in 2003. Yes, it was profitable on the short term, but as a company that had lived well on product innovation like GE had under Jones, squeezing a lot of profits out of it was not hard.

And not unlike Immelt today, the CEO of 3M, George Buckley, had to recover the business. The first thing he did was get rid of the Six Sigma program in the research labs. He noted that "one of the dangers of Six Sigma is that when you value sameness more than you value creativity, I think you potentially undermine the heart and soul of a company like 3M."

He is now funneling cash into what he calls the "core" areas of 3M technology – 45 in all, from abrasives to nanotechnology to flexible electronics. He sold the skin-care cream Aldara business that McNerney bought to harvest. Now 3M is plowing $1.5 billion back into 18 new plants and major expansions.

Home Depot also learned the lessons of Six Sigma, which was developed as a systematic way to improve quality, but has been used by CEOs as a way to only improve profits. Ironically, it was the second major contender for Welch's job -- Bob Nardelli – who introduced Six Sigma to Home Depot. He sent his Black Belts to all the stores to streamline the checkout process and to strategically place products. Again, a sea of Six Sigma paperwork to "define, measure, analyze, improve and control" drove employees nuts as they had less time to look after customers.

Profitability soared for the short term with layoffs and so did consumer sentiment. Home Depot dropped from first to worst among major retailers in the American Customer Satisfaction index in 2005. Nardelli's successor has now dialed back on Six Sigma and given store managers more leeway in making decisions on their own.

In fact, even Fortune magazine, which often heralded Welch as the manager of the century, is now taking a harder look at "appearance versus reality" in the business world. They state that of the 58 large companies that have announced the Six Sigma program, 91% have trailed the S&P since.

Fortunately for GE, Immelt (*not Nardelli or McNerney*) was chosen as the replacement for Welch. It is sarcastically said that Nardelli and McNerney successfully interviewed for other CEO jobs by listing the things one should not do.

Best of all, however, employees are no longer bombarded with Machiavellian misleading bumper sticker thinking and slogans. Instead, Immelt's business slogan is "GE imagination at work" as he pursues innovation and new markets, including the green ones -- and yes, he is hiring those competitive, competent American workers he said we were.

About once every three months, I have lunch with Mark Markovitz. Here are some of the questions we ponder over red wine and homemade pasta at Ferrari's, an Italian restaurant that now has seen its early GE immigrant workers become grey but still loyal.

> *What would GE be like today if during Welch's 20 year reign, 10% of its profits had been reinvested in research and product development?*

> *What would GE be like today if GE product businesses had not been harvested to turn GE into a financial conglomerate that nearly caused GE to crash in 2008?*

> *Japanese businesses just make annual reports to their stockholders and therefore do not have to constantly meet security analyst's expectations. Would GE and other businesses be better off with this system?*

What can be done about the corruption between some government officials and the upper one percenters which result in laws giving businesses extra incentives to move offshore?

How is employee morale affected when CEOs receive 500 times more --not the 15 to 20 it used to be -- than the average white and blue collar worker?

Leading the list of the "21 CEOS with Golden Parachutes of $100 Million or More" (www.GMIratings.com) is Welch at $417,361,902 when he retired in 2002 with the option of taking $96,000,000 or getting an annual pension of $9 million a year. How do retired employees, particularly some 600 just in Schenectady who get under $10,000 a year and have longer service than Welch, feel about that?

But most of all, a question we would like the greedy CEOs, bankers and corrupt politicians to answer this question:

What if somebody asked you to "work for a fair day's pay for a fair day's work?

Made in the USA
San Bernardino, CA
23 April 2018